1999 Toy Shop Annual

Foreword

My sister has begun collecting Barbie dolls — "collectors editions" — for my three-year-old niece, Abby. But here's the catch — Abby wants to actually *play* with them. GASP!

It's a Catch-22 for my sister. She wants to start a collection that may increase in value for her daughter, but she also wants to keep the toys in pristine condition. Try telling that to a three-year-old.

That's not an uncommon scenario to collectors, but it's a mentality from which some collectors refuse to break free.

I've had so much fun actually playing with toys this year that I'd like to recommend that diversion for all collectors. Sure, there are some toys I'm just content to put on the shelf and gaze at longingly (like my Fox Mulder action figure). But, then again, I don't collect solely for investment.

Take some time this year to actually pick up, unpackage and play with your toys. Your psyche will thank you, even if your co-workers look at you a little funny.

I know I'm a happier person because of my toys, and I thank Abby for helping me see the true meaning and carefree joy in what I do for a living. Not everyone is that lucky.

This annual directory is one I'm especially proud of. We have some especially timely stories about our hobby plus one of the most complete directories of clubs, dealers, manufacturers and toy shows nationwide. You're sure to find what you need in this one volume.

On page 39, you'll also read about collectit.net — Krause Publications' exciting new auction site. It's one of our first ventures into cyberspace collecting. Give it a shot to find the toys you've been looking for.

Toy Shop magazine celebrated 10 years in the toy collecting hobby in 1998. So to our readers we say thank you for sticking with us. We look forward to jumping into the next millennium with you.

Thanks to all the contributors who took time to provide interviews for this annual, especially the many national toy companies, dealers and collectors; associate editor Mike Jacquart; contributing writers Guy-Michael Grande, Christy Bando and Susan Mellish; Ross Hubbard, Kris Kandler and Chris Pritchard for cover photography and design; and the advertising staff of Toy Shop *magazine, an essential part of publishing this book.*

Sharon Korbeck

On the Cover

If you love toys like we do, you'll realize how exciting it is when the packages start to arrive for our cover photo shoot.

Almost every day, a new exciting toy crosses our desks. As editors and toy enthusiasts, it's our job to present our readers with the newest, the hottest and the best of the best.

It's hard to narrow down our favorites, but the toys depicted on our front and back covers represent some of the best of 1998. Here's an identification guide to the toys pictured.

Bandai: Deluxe Astro Megazord, Power Rangers series

Basic Fun: Lunch box key chain

Ertl: Case die-cast tractor, 1:16-scale

Galoob Toys: Ginger Spice doll, Spice Girls series

Gibson Greetings: Silly Slammers beanbag

Kenner/Hasbro: Chip Hazard figure, *Small Soldiers*; Omar Bradley, G.I. Joe collection; Winner's Circle Dale Jarrett Ford; Teletubbies figure; *Star Wars* Buddies

Mattel: NASCAR Barbie; Hot Wheels 30th anniversary Twin Mill; *Mulan* doll; *Armageddon* action figure

McFarlane Toys: *X-Files* figures; The Ogre figure, Dark Ages Spawn series

Parker Brothers: *Star Wars* Trivial Pursuit

Playing Mantis: *Lost in Space* model kit; KISS stock car

Playmates: Jean-Luc Picard figure, *Star Trek* series

Thinkway: Woody beanbag, *Toy Story* series

Toy Biz: Captain America figure, Famous Covers series; Lara Croft figure, Tomb Raider series

Toy Island: *Toonsylvania* figure

Trendmasters: Ultimate Godzilla

Ty: Rainbow the Chameleon Beanie Baby; Pinchers the Lobster Teenie Beanie Baby

Thanks to the companies listed above for supplying toys for the cover.

Table of Contents

1999 Toy Shop Annual Directory

Editor: Sharon Korbeck (korbecks@krause.com)
Associate Editor: Mike Jacquart (jacquartm@krause.com)
Contributing Writers: Christina Bando, Guy-Michael Grande, Susan Mellish
Cover design by Chris Pritchard
Visit us online at www.krause.com

Advertiser Index

Move Over, Godzilla!

Bugs May Take a Munch Out of Year-End Toy Sales

By Sharon Korbeck

Of all the movie features and creatures of 1998, it may not be Godzilla, *Small Soldiers* or a space robot that collectors will remember most, but perhaps a lithe little ant named Flik.

By early fall, 1998, retailers were already anticipating blockbuster sales of toys from Walt Disney's latest animated film, *A Bug's Life*, starring Flik the misfit ant.

"We're expecting it to be another *Toy Story*," said Doug Wysocki, who works in Inventory Control for Toys R Us. *A Bug's Life*, Wysocki said, is one of the first big children's movies of the year in terms of toy marketing.

A Bug's Life, starring the voices of dozens of stars, was slated to hit theaters in November and feature the

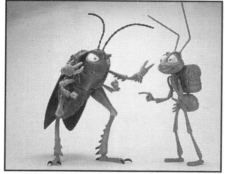

Retail stores were hoping for an insect landslide when Disney's A Bug's Life *hit stores in November. Mattel, Applause and Thinkway were major toy licensees.*

voices of dozens of celebrities. It was animated by Pixar, creators of the landmark *Toy Story* — a huge film and toy marketing success. Mattel and Thinkway Toys hold the key licenses; McDonald's will produce fast food toys.

But whether his little antennae charm buyers or not, little Flik crawled onto shelves after a year's worth

of bustling toy activity. Most of that activity — and a lot of the sluggishness — was directly related to movies.

The Lizard that Snored

Seems Sony Signatures got more mileage out of those Taco Bell ads with the chihuahua chanting "Here lee-zard, lee-zard" than it did with its *Godzilla* film itself. The movie was the talk of the American International Toy Fair, but Sony kept close wraps on the toys, allowing no sneak peaks before the film aired. But the movie opened and closed without much ado, and it

Love 'em or hate 'em, the Teletubbies were everywhere in 1998. Kids loved them, but will they still be around in 1999?

The year 1998 was a very good one for Todd McFarlane and his creative toys. His KISS Psycho Circus figures, like the one of Ace Frehley pictured at left, were eagerly sought by collectors for their finely detailed likenesses of the rock icons.

appeared the covert marketing wasn't worth all the advance buzz. And while some of the toys were exceptional — Toy Biz's remote-control Baby Godzilla remains a favorite — many languished on store shelves long after the film had been released.

Wysocki said he expected *Godzilla* toys to do better after the film is released on video.

Another much-anticipated summer, 1998, film was *Lost in Space*. Since vintage Lost in Space toys often fare well on the secondary market, collectors were anticipating a wellspring of new merchandise. Lost in Space toys, most made by Trendmasters, did well initially, according to Wysocki, but "sales leveled out. It dropped off quickly."

Faring marginally were toys from *Armageddon* and Disney's *Mulan*. It's unlikely either will do much on the secondary market, although Mattel's

The much-anticipated **The X-Files** *film opened with fanfare; collectors loved McFarlane Toys' action figures of Agents Scully and Mulder and the creepy film aliens.*

Armageddon action figures were substantial, detailed, heavy and pretty good likenesses of the movie's characters. *Small Soldiers* was a good late year contender for action figure sales.

What About *X-Files*?

Toy talk early in 1998 at Toy Fair centered around that hot license Todd McFarlane had secured —

the movie version of Fox's favorite, *The X-Files*. There was little doubt among collectors that McFarlane — known for value-packed, detail-laden, creatively-crafted action figures — could deliver.

McFarlane's action figures remained in demand months after the film's release.

But Mattel's X-Files Barbie and Ken was a different story. The two-pack gift set was pulled almost immediately after it hit store shelves due to a packaging problem. The very next day, those lucky enough to have purchased the very first $80 gift sets were able to sell them for $150 or more.

But once the corrected gift sets returned to shelves weeks later, the movie had died down and collectors were less than enthused.

Wysocki also felt the set didn't reach the right audience. "It didn't hit the Barbie market. The people who watch *The X-Files* aren't the people who collect Barbies," he said. He also believed that the

initial recall of the doll hurt retail sales.

Collector buzz was very high in February when Hasbro announced it would make a G.I. Joe figure in the likeness of Tom Hanks, who was starring in Steven Spielberg's World War II epic *Saving Private Ryan*. When Dreamworks officials deemed the project inappropriate (due to the graphic violence in the film), Hasbro scratched the figure.

Wysocki said Toys R Us did hear from collectors upset that the project was scrapped.

But G.I. Joe isn't the only guy in town anymore. Wysocki said 12-inch soldier figures by 21st Century Toys "are really butting heads with Hasbro" and giving good ol' Joe a run for his money. Wysocki said he's seen a "renewal in interest in G.I. Joe, 12-inch figures and 21st Century Toys products."

Wysocki said 21st Century figures have quite a few benefits, like offering lots of new styles for a lower price point. "It's not just collectors [who are buying], it's kids," he said.

More Action Figures

What other action figures interested buyers in 1998? Once again, McFarlane Toys ruled the shelves with its KISS Psycho Circus and Movie Maniacs. According to Wysocki, McFarlane's "Dark Ages Spawn has exploded."

The Toys R Us exclusive Munsters figures from Exclusive Toys "did

Another highly-anticipated film release — Lost in Space — failed at the box office. But some nice toys were made, nonetheless. We liked Trendmasters' replica of the vintage B-9 Robot.

extremely well," Wysocki said.

Kenner's Star Wars line experienced ups and downs. Wysocki said, "Star Wars is starting to move to an only collector item." With no new movie, no video release and no media tie-in, kids are moving away from Star Wars, Wysocki said, at least for now. The 1999 theatrical release of the *Star Wars* prequel film may re-energize Star Wars toys for kids and collectors.

High-end toys, however, like the $50 Han Solo and Tauntaun set, were excellent sellers. Another related winner was Parker Brothers' Star Wars Trivial Pursuit, one of Toys R Us's best-selling games of 1998.

Wysocki also expected Kenner's Planet of the Apes figures to sell well at the end of the year.

The Pink Aisle is Still Strong

It's always easy to find the Barbie aisle in any store. Just look for the pink.

Wysocki used the word "stable" to 1998 Barbie sales. "I see collectors coming back to Barbie," he added. "Some of the collectors' dolls

Providing a challenge to Hasbro's G.I. Joe was the military figures of 21st Century Toys. The company's Ultimate Soldier line received a lot of shelf space next to Hasbro's more well-known soldier.

This year's anticipated Christmas best-seller — at least at press time — was Mattel's Bounce Around Tigger. The plush Pooh character bounces, giggles and talks.

are going very well."

He cited NASCAR Barbie and Harley-Davidson Barbie, especially, as selling out immediately.

NBA Barbie, which came out in time for Christmas and the basketball season, was expected to sell well. Toys R Us 50th anniversary Barbie was expected to sell well despite a high price ($99.99).

In the past, as was the case with 1997 Happy Holidays Barbie, excessive supply left retailers with many unwanted dolls at the end of the season. This year, however, overproduction doesn't appear to be a problem; Wysocki said that may lead to swifter sales.

Among other best-selling 1998 dolls were Galoob's Spice Girls, particularly the Toys R Us exclusive five-piece set.

Hot Wheels Sets Records

"We'll have record sales for Hot Wheels this year," Wysocki said, even though Mattel's 30th anniversary Hot Wheels line didn't sell as well as Toys R Us thought it would.

A big problem, Wysocki said, was that stores couldn't get new cars in the line; the company kept shipping the same cars over and over.

Toys R Us will offer a complete assortment of the 30th anniversary cars, however, for collectors who couldn't find them all.

Racing Champions' Gold series was probably the store's best die-cast seller, Wysocki said.

What's Left?

Since major retailers don't sell Ty Beanie Babies, what fills the gap? Other plush products seem to be doing just as swift business. Wysocki said Blues Clues, a Nickelodeon television favorite, plush sold well. But Playskool's beanbag versions of the Teletubbies were consistently selling out in one day.

What will be this year's Tickle Me Elmo? As of October, Wysocki said, "Bounce Around Tigger is the closest thing we have." The Mattel product is a plush version of Tigger that, when pushed, talks, giggles and bounces up and down. It retails for $30.

Lots of collectors were happy to see likenesses of their favorite TV ghouls — The Munsters — made by Exclusive Toy Company. The company made many other figures of classic characters in 1998 including characters from James Bond films, **The Dukes of Hazzard** *and* **Babylon 5.**

Best of the Batch

Toy Shop Picks its Favorite 1998 Toys

Cream of the crop. Best of the best. Top of the heap.

OK, you get what we mean. No matter what the retailers, kids or toy companies say, everyone has their own opinion of the greatest toys of the year.

We've selected 12 of our favorites to showcase here.

We've seen a lot of toys this past year — some great for collecting investment, some great for kids, some just plain great. The list below is a little bit of each. Some certainly have better collecting potential. But all are certain to charm recipients this holiday season. Plus, almost everything on the list costs less than $100.

Happy shopping!

1. The Island of Misfit Toys Plush, Stuffins/CVS. Stuffins, a small company in New Jersey tops our list this year thanks to the popularity of the perennial animated TV favorite, *Rudolph the Red-Nosed Reindeer.* Remember Herbie — the elf who wanted to be a dentist? Or what about the misfit Charlie-in-the-Box? Or Burl Ives as the

voice of the roly-poly Snowman? These memorable characters — and nine others — are being sold as 12-inch and 6-inch plush — only through the East Coast drugstore chain, CVS. That's the catch — these extremely well-executed plush are only available at the stores, not via mail order. So plan your Boston trip now, or, like most of our *Toy Shop* staff has already done, cozy up to those long-lost East Coast relatives. You'll be glad you did. ($5.99 each/6-inch; $12.99/12-inch).

2. *A Bug's Life* Deluxe Talking Flik, Mattel. *Antz* may have beat Disney's film to the theaters, but it will definitely be *A Bug's Life* that will clean up store shelves. With several major licensees, there are literally colonies full of toys to choose from. We especially like the big, bold, blue Flik — a talking ant that speaks of "kicking some grasshopper abdomen" among other delightful movie phrases. ($30).

3. Bounce Around Tigger, Mattel. Finally, a Tigger that bounces as well as his animated character always said he could.

No doubt about it — Tigger is this year's Tickle Me Elmo. He's plush, he bounces and even erupts with his trademark giggle. Too cute for words. Enough said. Buy two now, and make them bounce and giggle in stereo. ($30).

4. *Star Wars* Buddies, Kenner. Hey, you don't have to be a *Star Wars* fan to think Yoda and Max Rebo are cute. Beanbags, of course, were the craze of 1998. *Star Wars* is the craze of the past 20 years with collectors. Put them together, and, well, the result is spectacular (except, what happened with Jabba? He looks like he's made out of recycled parachute pants.). Overall, the line is classic. ($5-$7 each).

5. Universal Monsters figures, Hasbro. Not the most articulated (or articulate) action figures, but definitely well-executed and packaged. We think the new year could be haunting thanks to these long-dead revivals. ($20-$25).

6. Evel Knievel Stunt Cycle, Playing Mantis. He's rubbery and ready to roll! After a long hiatus in toy land, Evel Knievel has been regenerated . . . and so has his stunt cycle. True to Ideal's original, this set will have executives revving up their engines and racing these childhood classics down company corridors.

7. Happy Holidays Barbie, Mattel. Barbie's back (and it may be here last year as a holiday issue) and it's about time . . .

we were getting tired of seeing surplus 1997 dolls littering shelves. Luckily, Mattel's updated her gown, dressing her in stunning black velvet and rhinestones rather than last year's tacky lace atrocity. The African-American version this year is especially striking. Take a chance on Barbie again; after all, she does turn 40 in 1999. ($40).

8. Movie Maniacs, McFarlane Toys. If Leatherface showed up at your house with a chainsaw what would you do? We'd go out and buy

the rest of Todd McFarlane's creepy Movie Maniacs line to keep him company. Scream flicks have been the rage for several years now, so it's no wonder these film classics have been immortalized by the reigning king of action figures.

9. Star Wars Trivial Pursuit, Parker Brothers. Maybe only die-hard *Star Wars* fans can answer all the questions, but this is one special edition game not lost on the rest of us. As an iconic property, *Star Wars* can't be beat. And it might be good to play a few rounds to get geared up for next year's Star Wars prequel.

10. Kiddie Car Classics, Hallmark. Technically, these aren't toys, but they are just too well-done to ignore. (Plus, they're great display pieces; we have Teenie Beanies at the wheel of ours). Hefty, shiny, authentic, movable . . . you run out of adjectives for these stylish replicas of our favorite vintage vehicles. ($50+).

11. Silly Slammers, Gibson Greetings. "You're bothering me." "You're fired." "Whatever!" The phrases uttered by slamming these bean-bags are hilarious and original — making these inexpensive toys office favorites this year. Buy a bunch — there's sure to be a phrase

appropriate for your favorite family member or co-worker. ($5).

12. Hot Wheels Legends Twin Mill, Mattel. Don't miss this opportunity to own the 30th anniversary Twin Mill, one of Mattel's earliest Hot Wheels. The two-car Legends set features a totally hot design, retro styling, flame-tread wheels and redline tires. It's our fave of the anniversary offerings this year. ($120).

In researching this story, Toy Shop*'s editorial staff spent hours bouncing Tigger, racing Evel Knievel and hiding Frankenstein under their boss's desk. Now, if we could just get Freddy Krueger to hunt down those blasted Teletubbies . . .*

POP CULTURE CLASSICS
commercials from the 50's & 60's

$9.95 each tape plus $4.95 shipping & handling $.85 each additional

CATALOGUE $15.00 History of Toys, Space Program, Westerns, TV Shows. Fully illustrated and plenty of surprises. Video's for sale– Classic Commercials, TV Variety Shows, Western, Sci Fi, Cartoons, Kids TV Shows from the GOLDEN AGE OF TELEVISION.

CEREAL COMMERCIALS – Kellogg's, Trix, Tony The Tiger, Superman Gang, General Mills, Lucky Charms, Pebbles, Rice & Wheat Honeys with Buffalo Bee, Maypo & more.

TELEVISION TOYS – Mr. Potato Head, Mr. Machine, Trick Trac, Shooten Shell Guns, Dick Tracy Two Way Radio, Shirley Temple, Astrobase, Mystery Date, Barbie, Ken & more

ANIMATION COMMERCIALS – Twinkles, Hamms Beer Bear, Fruit Stripe Gum, Tang with Bugs and Daffy, Fred Flintstones and Barney for Winston, Ajax and the Elves & more.

CAR COMMERCIALS – Edsel introduced by Bing Crosby in 1958, Ford demonstrates the future of Power steering, 64 Pontiac Grand Prix, 57 Chevy, Thunderbird, Buick, Hertz puts you in the drivers seat.

DOLL COMMERCIALS – Betsy Wetsy, Chatty Cathy, Tuesday Taylor, Tiny Tears, Revlon, Playtex, Barbie.

BEST OF COMMERCIALS ONE – Brylcreme, Chunky, Bosco, Double Mint Twins, Bonomo's Turkish Taffy

BEST OF COMMERCIALS TWO – Lustre Crème, Yodels, Cracker Jacks, Bond Bread, Wildroot Crème, Mounds

Make Checks Payable to:
NORMAN CHARLES CORP.

220 West 71st Street #54
New York, NY 10023

Website: TELEVISIONTOYS.

Phone: 800-442-7055
Fax: 212-595-0189
Email: VIDRES@AOL.COM

THE COLLECTOR's EDITION
Your favorite commercials:
Cereal, TOYS, Animation, CARS, Dolls
Best of One, Best of Two, Best of Three
SPORTS, Cigarettes

Each tape 30 minutes (SP) speed

SEND FOR: FREE NEWSLETTER

Major Acquisitions Abound
Hasbro Buys Galoob; Mattel Buys Pleasant

By Mike Jacquart

Whether it's Chrysler merging with Mercedes-Benz or paper giants Fort Howard and James River joining forces, it's clear in the business world that the big just keep getting bigger.

That was true in the toy world as well in 1998 — a year that saw the multi-billion-dollar Hasbro company, the world's second-largest toy company, purchase Galoob Toys for $220 million. Toy giant Mattel, the world's largest toy manufacturer, bought doll manufacturer, the Pleasant Company, for $700 million.

Hasbro Buys Galoob

With Galoob's popular *Star Wars* Micro Machines coming under its umbrella, the acquisition gave Hasbro a near monopoly on the solid-selling *Star Wars* license — especially noteworthy in light of the new movie (and toys) that will be released in 1999.

Wayne Charness, senior vice president of corporate communications with Hasbro, said he was not at liberty to discuss specific terms of the sale until the purchase is voted on by share-

holders (which had not occurred at press time), but he did note that "Galoob helps give us what we feel is a global force in the market."

Todd Lustgarten, Galoob's director of marketing for Micro Machines, said until the deal is formalized, it will be "business as usual" at Galoob. Lustgarten added he's excited about the increased buying power and capital Hasbro will be able to provide to Galoob toy lines.

Mattel Purchases Pleasant

Mattel's deal gave the company control of the No. 1 and No. 2 toy

brands for girls. Sales of Pleasant Company's American Girl dolls, books and clothes — which help educate girls about American history — reached $300 million in 1997. Mattel sold $1.9 billion worth of Barbies and accessories the same year.

Ironically, Pleasant Company founder Pleasant Rowland had previously blasted traditional Barbies as "brainless." The move also gave Mattel access to the Middleton, Wis.-based Pleasant's direct-mail facilities. The Pleasant Company is one of the 20 largest direct-mail firms in the country.

Other Buyouts

But acquisitions in 1998 didn't end there. Even prior to announcement of the

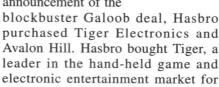

blockbuster Galoob deal, Hasbro purchased Tiger Electronics and Avalon Hill. Hasbro bought Tiger, a leader in the hand-held game and electronic entertainment market for $335 million.

Some of Tiger's top sellers include Giga Pets, Game Com and Laser Tag. The purchase was announced in February.

Hasbro also bought the Avalon Hill game line from Monarch Avalon for $6 million and computer game producer Microprose, Inc., for approximately $70 million. As part of the deal, Hasbro received all of Avalon Hill's inventory. Avalon Hill was to continue to sell its Civilization board game and its computer games until the end of the year — and its 1830 board game indefinitely.

Poof Products, a leading manufacturer of foam toys, acquired James Industries, the original manufacturer of Slinky toys. The product line purchase included the original Slinky,

several plastic Slinkys and the Slinky Dog that appeared in *Toy Story*.

Products of the Plymouth, Mich.-based Poof include foam flying toys ranging from airplanes and helicopters to a wide variety of sport balls.

Sales Down at Toys R Us

But while acquisitions were good news for large companies — as well as often being good news for the firms bought out because stock value often increases — all was not well in the toy world in 1998.

Engaged in a battle for market share with Wal-Mart, Target and other large retailers, Toys R Us announced in September a restructuring that was expected to lay off up to 3,000 workers, close 90 stores and reformat current ones to add children's clothing and electronic items like cell phones.

Why the problems? According to Sean McGowan, an industry analyst with Gerard Klauer Mattison in New York City, the toy giant has an image problem with teenagers, who will not venture into the retail stores.

The problems at Toys R Us also reflect an industry-wide trend in which retail sales were up less than one percent over 1997, according to the NPD Group, a New York-based marketing consulting organization. Even factoring in the holiday shopping season, NPD predicted just a two percent hike overall in 1998.

The stagnant cash registers are at least partly due to the fact that kids' inter-

est in toys is peaking one or two years earlier than a decade ago, said John Eyler, CEO of FAO Schwarz.

In the meantime, toymakers are aggressively pursuing new venues like Mattel's customized online Barbies — as well as keeping their fingers crossed that a super-selling toy line will emerge to rejuvenate sales.

That's Not Mint!

Laments about Toy Grading

By Sharon Korbeck

Has the word "Mint" lost its meaning? Far too often these days, people advertise Mint condition toys that are, actually, far from that condition.

Maybe it's ignorance, possibly it's just wishful thinking. But whatever the reason, dealers past and present have continued to overestimate the condition of their toys. And many unassuming, or unknowledgeable, collectors have fallen victim.

At one show, a dealer had a Munsters board game priced at $150 (its book value in Mint condition). When asked why the price was so high, the dealer responded, "That's what it books for in the price guides."

What the dealer failed to take into consideration, however, was that the board game he was offering was not in Mint condition — it had a water-damaged cover, broken side apron and several pieces missing. That condition probably warranted that the dealer take at least 50 percent off the game's price.

But offering lesser condition toys for top prices is far too prevalent in the hobby — and it will continue unless more people become educated about how to grade toys reasonably.

Grading scales vary from dealer to dealer, publication to publication and collector to collector. Even geography plays a role.

The reasons for buying toys may also determine what a collector is willing to pay. Some people seek toys in the best possible condition for investment purposes; others may seek the same toy in any condition for nostalgic reasons.

While grading is an imperfect science, here are some tips to keep in mind.

1. Packaging. Most toys originally packaged on cards or in boxes command higher prices in their new, pristine condition even though they may be more beautiful outside the package.

Questions to ask yourself about packaging include: Does the box exhibit shelf wear? Are the edges worn or lightened in color? Does a blister card have a crease or scratches? Is the plastic bubble perfect? Even subtle imperfections are a big factor in determining what a hard core collector will pay.

2. Material. What is the toy made of and how sturdy is that material? For example, paper items (like boxes, paper dolls and coloring books) will likely exhibit some wear over time. Some wear on these items may be acceptable to collectors; major wear, however, may not. Cast-iron toys, which should have held up better over time, should be judged differently than toys made of less-sturdy construction.

3. Age of Item. When was the toy originally made? Late 1800s? 1950s? 1990s? A collector buying a toy from the early 1900s may be willing to accept a lesser-condition item since so few examples may still exist (and the toy has had to survive over 90 years). But toys from the past 20 years often must be in much better condition to entice buyers, unless the toy is ultra-rare or desirable.

4. Appearance and Function. These two factors are often vital to the collector, and they go hand in hand. For example, a vintage Fisher-Price pull toy may look great, but does it work as it was originally intended? Some collectors, however, may not care whether or not a toy is operational as long as it looks good for display purposes.

Toy Shop has adopted the following grading descriptions. Remember, however, no single grading system will apply to all toys. Some descriptions may not apply to certain toys depending on how they are categorized, when they were produced, how they were originally packaged and how they are collected today.

MIB / MIP (Mint in Box, Mint in Package): Just like new in original package. The box may have been opened, but inside packages remain unopened. New toys in factory-sealed boxes may command higher prices. A comparable rating used in other guides may be "C-10."

MNB / MNP (Mint no Box, Mint no Package): This is also known as "loose" condition. A toy in Mint condition but not in its original package.

NM (Near Mint): A toy that looks new in overall appearance, but exhibits very minor wear and does not have the original box. An exception would be a toy that comes in kit form. A kit in Near Mint condition would be expected to have the original box, but the box may display some wear.

EX (Excellent): A toy that is complete and has been played with. Signs of minor wear may be evident, but the toy is very clean and well cared for.

GD (Good): A toy that obviously has been played with and shows general wear overall. Paint chipping is readily apparent. In metal toys, some minor rust may be evident. In sets, some minor pieces may be missing.

Take a look at our grading system, and compare it with others you may encounter. The basic tenets of condition and appearance apply, although perhaps at different levels. For example, Richard O'Brien's *Collecting Toys* books use a C6, C8, C10 grading system, which compares fairly closely to Good, Excellent and Mint conditions.

Decide what type of collector you want to be — the level of money you are willing to spend, the amount of return (if any) you hope to reap from your investments and the amount of time you will devote to your hobby.

Your purpose and mission as a toy collector will figure into how much you are willing to spend.

So, learn the jargon, develop a sharp eye for subtleties, compare grading scales and decide what a toy is worth to you. You can turn the subjective field of toy collecting into a rewarding objective for yourself.

The Big B's of 1998:
Barbies and Beanie Babies

By Susan Mellish

The hottest toys of the past few years can be boiled down to one letter — "B."

Barbie and Beanie Babies.

Think about it. Both of these collectibles are easily recognizable by just about anyone walking the planet. I'd bet the average person knows more about Barbie or Beanie Babies than they do about American history or their local tax laws.

But even though they are two extremely popular collectibles, Barbie and Beanie Babies hold very unique, and different, places in the arena of collectibles.

Barbie wasn't created as a collectible, but rather as a plaything. Initially, Barbie creator Ruth Handler recognized a need for a fashion doll that little girls could play with. Barbie was an immediate success.

That was in 1959. Jump ahead 40 years, and Barbie is still as popular as ever with little girls, but she has also made her mark as one of the most collectible items available on the market; attractive to collectors of both genders and all ages.

Aided by Ty's creation of an official Beanie Babies Collectors Club, the popularity and value of the beanbag toys soared in 1998.

It is because of this changeability, this being in tune with the times that Barbie continues to appeal to collectors today. Not only does Barbie continue to be a quality product, the makers at Mattel have made a commitment to collectors to offer better items for as long as the demand exists.

of a collecting phenomenon came into being.

Beanie collectors are banking on these little beanbags stuffed with PVC pellets. Still, they have a very short history compared to Barbie, and it is uncertain how long these $5 toys can keep commanding inflated prices.

Now that 1998 is over, let's take a look back and see how the two "big Bs" in collecting fared. What was the most exciting incident in each field this year? How do collectors and dealers view the market today and where do they think it is going?

Brian Devany, president of Beanie Source International, sees this collectibles area as a very viable field. He has such faith in the continued popularity of Beanie Babies that he recently quit his job in order to work full-time as a secondary market source for Beanie Babies. He specializes in wholesale supply to dealers and retail sale of the upper echelon retired Beanies. Devany is also marketing a business plan to help others do the same.

Devany got into the business side of Beanies in March, 1997, about six months before this phenomenon really exploded, and he still feels the popularity of Beanie Babies will continue.

"As long as collectors outnumber the dealers, and collectors well outnumber dealers, this area will continue to grow," Devany, who presently has a quarter million dollar inventory of Beanies, explains.

Is Devany concerned that Beanies will drop in popularity? "I don't worry about having a lot of inventory. Selling Beanies is one of the few businesses where the inventory might actually increase in value the longer it sits. Right now, having Beanie Babies

---

"Right now, having Beanie Babies is better than having a stock certificate."

It was because of her popularity that production of this product continued and evolved as our culture evolved. It is relatively easy to date a Barbie or a Barbie fashion if one knows anything about fashion history or what was "in" at the time.

Unlike Barbie, however, Beanie Babies, first produced in 1993, took a little while to catch on. The plush beanbags didn't really ignite a firestorm of activity until manufacturer Ty "retired" the first Beanies in 1996. Madness took over, and nothing short

is better than having a stock certificate," he said.

In Devany's opinion, demand for Beanies runs in waves. For example, in October, 1997, demand was very high. That demand continued through Christmas until well into March, 1998. From mid-April to the end of July, demand tailed off a bit, prices dropped and the supply got big, though the sales for the secondary market stayed strong. Devany stated that with back to school shopping, August saw prices firming up again, and he expected the prices (on the secondary market) to rise again in September (just before this issue went to press).

The most exciting happening in the field of Beanie Babies last year, Devany said, was the Teenie Beanie Baby promotion by McDonald's in May; it was a mass buying frenzy that resulted in 200 million Teenies being distributed in less than two weeks.

"It brought many, many new collectors into the field," he pointed out. "These new collectors then began purchasing the current Beanies available in stores. Then if they got all of those, they started searching the secondary market for more," he explained.

He feels that part of the fun of collecting Beanies is the hunt. Devany states that before collectors know it, they have run their credit cards up to the limit in order to purchase the hard-to-find Beanies.

Devany also noted that the Beanie Baby name is now as recognizable as baseball and apple pie. "Everyone knows what a Beanie Baby is, and I've also noticed that people don't feel that spending hundreds or thousands of dollars on a Beanie Baby is such a crazy idea anymore."

"If anything will hurt this collecting field, it will be the fakes," states

Mattel's Barbie and Ken as Scully and Mulder of **The X-Files** *received a lot of attention from collectors. Initial shipments of the $80 set were pulled from store shelves due to a packaging error. Prices immediately shot up within days; later in the year, however, the set was readily available on many store shelves at retail price.*

Devany. "If they are not controlled, this market will crumble from the top down."

Devany also sees the need for some organization in the secondary market. "Something needs to happen to organize the secondary market dealers, to legitimize this business so collectors feel secure knowing they are buying from a reputable dealer."

"I'd have to say the NASCAR Barbie was the most popular doll of 1998."

Still, Devany feels that Beanie Babies have to reach that first plateau — the five-year mark from the date of the first retirements before everyone can say that Beanie Babies have longevity as a collectible.

"The consensus with many people is that Beanie Babies are still just a fad. We have to overcome that five-year hurdle, and until we reach that point, I'm not prepared to say that Beanie Babies are a forever thing, but I really do not see their popularity going downhill anytime soon!"

Beanbag Bandwagon

But since Beanie Babies have become so popular, dozens of other companies have begun making beanbag toys — hoping to jump on the beanbag bandwagon.

Shawn Brecka, author of *The Beanie Family Album and Collector's Guide* (Antique Trader Publications), says Disney and Warner Brothers are also offering beanbag products that are, in her opinion, more interesting and of higher quality than Beanie Babies.

Ty may have had the jump on any other manufacturer of beanbag toys, but Brecka sees Disney and Warner Brothers not only catching up, but possibly surpassing Ty's Beanies where longevity is concerned.

Disney and Warner Brothers beanbags carry the added benefit of name and character recognition. Everyone knows who Winnie the Pooh is. People in their 30s and 40s grew up with Scooby Doo.

"Both of these companies also offer collectors a challenge by frequently retiring products and bringing out new versions," Brecka points out. "There is that added challenge for those collecting Disney beanbags because different beanbags are offered at The Disney Store as compared to the theme park gift shops and the Disney catalog," Brecka adds. "Collectors really have to stay on their toes!"

Where Ty Beanie Babies are

concerned, Brecka states, "I do not see their popularity waning anytime soon. Their market is stabilized, but you're not going to see the growth its seen before. Beanie Babies are no longer the only popular beanbag toy on the market, and they [Ty] need to take that into consideration, not only in deciding how often to bring out new Beanies, but also in how the Beanies will be distributed for the public to purchase."

One other major factor concerning Ty Beanie Babies would have to be their association with professional sports in 1998, mainly baseball. Whether giving away Beanies and their collectible cards as promotional items helped increase the numbers of Beanie Babies collectors is debatable.

HOT DOLL — This Harley-Davidson Barbie doll was one of the collector favorites of 1998. She was one of two dolls made honoring the motorcycle giant. Secondary market prices for this doll easily rose to more than $200 immediately after its release on the primary market.

Whether their being given away at major league baseball events helped sell tickets is undeniable. Games that, in other instances, would have seen spartan attendance at best, quickly sold out when it became known that a Beanie would be given away to the first wave of children under age 14 who entered the ballpark. As we head into 1999, the National Football League and the National Hockey League are lining up to use Beanies to also fill their stands.

A story about Ty Beanie Babies in 1998 cannot be complete without including how this product has been used to raise money for charities of all kinds. The Diana, Princess of Wales, Memorial Fund aside, local charities; non-profit organizations; schools and children's sports programs all saw an increase in their coffers if Ty Beanie Babies were part of their fundraising raffle or auction.

Don't Forget Barbie!

The other big "B" of 1998 was Barbie — a doll that needs no introduction. And though 1998 Barbies kept collectors "wowing," 1999 is anticipated to be a banner year with the American icon doll turning 40.

Sandi Holder of Sandi Holder's Doll Attic in Union City, Calif., feels that the NASCAR Barbie, celebrating the 50th anniversary of the stock car racing association, was very, very popular in 1998.

"I also think the second edition Harley-Davidson Barbie was a winner," Holder stated. "The demand for the first edition in 1997 was also huge, which I don't think Mattel was prepared for and many collectors missed out because scalpers scooped up the small amount of dolls available. This year Mattel made more dolls so collectors wouldn't run into that problem."

The Midge Gift Set, a Toys R Us exclusive, truly caters to collectors, Holder said. "Mattel really made the Midge set authentic, and collectors appreciate this," she explained.

Twist and Turn Smasharoo is another example of this. Exclusive to doll shops and FAO Schwarz, this doll

THE PRICE MAY NOT REMAIN THE SAME — Ty's Princess Beanie Baby is one example of deflation in the market. Even though the beanbags sold like mad throughout 1998, the commemorative Princess bear — once offered and bought at astounding prices of $200 or more — sunk down to the more realistic $5-$20 by the end of 1998, once more were available to the general public.

features open packaging, "just like the original dolls," Holder said.

Another doll offered in 1998 that Holder felt was well received by collectors included Bob Mackie's Fantasy Goddess of Asia.

Overall, Holder sees Barbie as being a very solid market. "Collectors are very appreciative of Mattel's increased attention to detail and fashion and that collector dolls are again being limited somewhat in production," Holder said.

Ask Marl Davidson of Marl & B of Bradenton, Fla., which Barbies she thought were hot sellers in 1998, and you'd better be prepared for a list.

"Right off the top I'd have to say the NASCAR Barbie was the most popular doll of 1998. The X-Files set was also very hot, and everyone loves the Romeo and Juliet set," Davidson stated.

"Oh, and the Twist and Turn

Smasharoo Barbie which is a copy of the 1996 Twist and Turn Barbie is absolutely great."

Davidson sees the Barbie market as being very steady, though maybe it slowed down some in 1998. "I expect the area to take a big surge in 1999 as Barbie celebrates her 40th birthday next year," Davidson points out.

She also sees Barbie's 40th anniversary bringing in new collectors, but she still feels people should collect what they like the best. "Maybe start off by collecting a series you like," suggests Davidson, "and then let it build from there."

Davidson adds, "Mattel is getting better and better with face sculpting, capturing the likenesses of special dolls. The designing of the clothes, the boxes, everything has improved and Mattel should know that collectors really appreciate that the company is listening and giving them what they want. That's why collecting Barbie will continue — because Mattel has listened."

"I love hearing that collectors know we care," states Suzanne Schlundt, senior manager of collector relations for Mattel. "That's what Mattel strove to do, and we delivered in 1998. We carefully assess every element of each doll we make in the collector lines being very picky about which face sculpt we use, which hairstyle, face paint and making sure the quality of packaging meets the high standards set by those who collect

Barbie."

What did Schlundt see as the hot Barbie item in 1998? "That's a tough one, but the Dolls of the World line and the ballerina series were very well received," Schlundt said.

The Barbie and the Tale of Peter Rabbit (Keepsake Series) and Angel of Joy (Timeless Sentiments) dolls

TEENIE BUT MIGHTY — McDonald's 1998 Teenie Beanie Babies promotion drew crowds to the fast food restaurant for the tiny toys. Some stores sold out within a few days of the promotion's start. Complete sets are especially sought by collectors. BELOW LEFT: It was a very good year for collector edition Barbie dolls, especially those created by fashion designers like Nolan Miller. His Barbie draped in lacy black was a high-end collector favorite in 1998.

were also popular. Schlundt echoed Holder and Davidson in stating that NASCAR Barbie was huge.

"But this makes sense when you look at the logistics of NASCAR," states Schlundt. NASCAR racing is the number one spectator sport in the nation, and a major portion of those fans are female.

"Mattel sponsors Kyle Petty, and all his sponsors agreed to our

putting their logos on Barbie's outfit, so the doll is really special to NASCAR fans," Schlundt points out.

Other popular Barbies for 1998 according to Schlundt? Again, the X-Files set and Harley-Davidson Barbie (Barbie in an edgy, hip, leather-clad look) were big. And while Bob Mackie's yearly edition (Fantasy Goddess of Asia) is always highly anticipated, another famous Hollywood fashion designer, Nolan Miller, dressed Barbie in 1998. And can you believe 1998 marked the 11th year of the Happy Holidays Barbie series? Time flies!

How does Mattel continue to keep Barbie collectors happy after so many years?

"We try to launch new Barbies all year long, more than 65 collector and limited-edition Barbies this year alone. Mattel has a strong commitment to collectors and we know how important quality is," Schlundt said.

She added, "For the past 40 years, Barbie has remained the world's favorite fashion doll, and I don't anticipate that changing anytime soon. With her 40th anniversary next year, we expect to attract new collectors, and an influx of new collectors always adds more excitement. Because of Mattel's commitment, 1998 was the best year ever, but wait until you see what we have to offer in 1999!"

And so 1998 was big for both Barbie and Beanie Babies. Barbie continues to grow old gracefully. No one even feels the need to comment if Barbie will be collectible five years from now.

Beanie Babies, a true infant in the collectibles field, still have to prove their longevity. With all this in mind, 1999 should be a very interesting year!

Brian Devany can be reached via e-mail at devany@inetport.com. Sandi Holder can be reached at 2491 Regal Dr., Union City, CA 94587, 510-489-0221 or via e-mail at SandiHB4U@aol.com. Marl Davidson can be reached at 10301 Braden Run, Bradenton, FL 34202, 941-751-6275.

Auction Action 1998:

Mechanical Banks Among Record-Setters

By Christina Bando

From vintage dolls to teddy bears to pedal tractors, the toys of yesterday brought record prices at auction in 1998.

Despite growing interest in new

The Zig Zag Bank — the only known example of this unusually-styled bank — sold for $189,500.

toys, such as *X-Files* dolls and figures, Beanie Babies and the Hot Wheels Treasure Hunt Series, 1998 auctions showed many collectors' willingness to hold onto the past.

The $426,000 Bank

The highlights of 1998 auctions were the mechanical banks. Bill Bertoia Auctions made auction history with total sales of over $4 million.

The top selling bank from the Stanley Sax collection was the Old Woman in the Shoe. Manufactured by W.S. Reed Co. of Leominster, Mass. in 1883, the rare bank commanded $426,000, the highest price ever paid for a mechanical bank in auction history.

The politically incorrect Darkey and Watermelon bank (also known as the Foot Ball Bank), followed close behind, realizing $354,500. The bank was produced the same year the first American Football League started and reflected American prejudices of the time.

Mechanical bank collectors were treated to over 251 antique banks, all presented in alphabetical order. Other top selling banks included the Bank Teller, the Freedman's Bank and Professor Pug Frog's Great Bicycle Feat.

The Bank Teller, a rare, highly-sought J&E Stevens bank, sold for $96,000 due to its "one of only six known" status. The Freedman's Bank, representing a post Civil War financial institution formed to help freed slaves, realized $321,500. The Bicycle Feat bank sold for a whopping $96,000.

The final item sold, the Zig Zag Bank, had an estimated value of $100,000-$125,000. Composed of cast iron, tin, papier mache and cloth, and patented to Moses Newman and George Henry Bennett in 1889, it sold

ABOVE: The Old Woman in the Shoe sold for $426,000 at a Bill Bertoia auction. It was the highest price ever paid for a mechanical bank. BELOW: The Mikado Bank realized $123,500.

for $189,000.

Hold the Fort Seven Hole Bank, a cast-iron bank from 1877, commanded top prices at an auction by Henry/Peirce Auctioneers. The bank featured a cannon and sold for $7,500.

The banks were taken from a 30-year collection of Tam and Bob Watkins. Other big sellers included a Palace bank that took second place at $5,200, a Lighthouse bank from 1891 that realized $2,200 and a J&E Stevens General Butler still bank that brought $2,100.

Other unusual cast-iron banks were offered with an ironic twist: they depicted banks. Among these, the Fidelity Trust Vault bank took the top slot at $320, followed by a City Bank at $190.

Pedal Cars Were Popular

From pedal cars to model trains to ships, 1998 was a great auction year for transportation toys.

Over two dozen pedal cars were sold at Dana Mecum Auctions, many made by AMF and Murray. On the top of the list was a 1955 Murray Dipside Fire Truck with ladder, which sold for $1,800. A 1948 Pontiac fire truck realized $1,100, while a three-wheeler 1960s Murray Trike sold for $1,100.

Perhaps the largest pedal toy sale was held by Aumann Auctions, spanning over three days with items ranging from farm tractors to petroleum collectibles.

A Coffin Block John Deere A sold for an incredible $25,000 at auction, which is believed to be one of the highest prices ever paid for a pedal tractor.

The second highest bid came from a Cast in Rear Hub Farmall Tall M at $13,500.

Ford pedal tractors did well, selling from $4,000 to $4,250, two of which were missing parts. The Ford Commander brought in $1,700.

Streamlined, sleek pedal cars and airplanes were the order of the day at Noel Barrett's antique toy auction. A pressed steel Steelcraft Streamliner pedal car sold for $2,530, while an original Steelcraft Spirit of America pedal airplane realized $9,900.

Pedal cars and planes were not the only popular transportation toys in 1998. Toy cars, planes and trains also fared well at auction.

Bugatti Model Brings $14,000

A hand-crafted, half-scale Riviera Blue Bugatti Type 59 model was orig-

ABOVE: The politically incorrect — by today's standards — Darkey and Watermelon — brought $354,500 at a Bill Bertoia sale. BELOW: Another important piece of Black Americana, the Freedman's Bank, realized $321,500. The bank was named after a post Civil War financial institution formed to help newly-freed slaves.

inally conceived as a gift for the engineer's son. However, the engineer spent more than 1,500 hours on the creation, using factory drawings for reference and extensive conversations with the Bugatti Trust.

The model powered by a 24-volt motor, took the top spot at a Brooks auction with a $14,915 bid.

Also included in the auction was a single-seat Fiat Grand Prix and Landspeed Record car that sold for $956; an Opel child's car for $3,441; and a Whitanco mechanical tank at $726.

Model cars and transportation toys topped the best-seller lists in other auctions as well. A Bing race car from the early 1900s brought $44,000 at a

A half-scale Riviera-blue Bugatti Type 59 model brought $14,915 at a Brooks auction. This superb model is powered by a 24-volt motor and features crafted wheels, suspension and chassis, with period dashboard and driving arrangement.

This Opel child's car brought $3,441 at the Brooks sale. It has original lamps, linkage and weathered maroon paint but lacks wheels.

Noel Barrett Auction, while a German Swan Neck Sleigh sold for $20,900.

Two Carette limousines brought in $22,000 apiece. One of them was well over its presale estimate of $5,000-$6,000.

A rare JEP Renault Tourer managed to top the sales in a Bertoia auction at an eye-popping $39,600. The Bertoia auction gave a wide selection of prewar vehicles, including a Rossignol tin turbine car from 1900, which sold for $22,000; a 12-inch Gorden Bennet race car at $20,900; a German tin lithographed taxi at $6,050; and a Smith-Miller Sibley's moving van at $6,875.

Right alongside the model car in terms of popularity were the model trains of 1998's auctions. A "Princess Elizabeth" Pacific locomotive and tender by Bassett-Lowke brought in

Auction goers were stunned when this 1:85-scale wood model of the ship Ellen May sold for $2,200 at a Garth's auction.

$5,736 at a Brooks auction.

The Brooks auction also included a small series model of the most famous British Pacific locomotive, the "Flying Scotsman," which sold for $4,780. The model was circa 1935 and extremely rare. "Sir Alexander Henderson," circa 1925, sold for $3,633 while a coal-fire 3-1/2 inch gauge Pacific locomotive sold for $1,912.

Finally, an aluminum three-car, electric Burlington Zephyr set ranked right up there with a selling price of $4,780.

Model trains dominated the top spots of Garth's Auction as well. The highest bid went to a Samhongsa Southern Pacific 4-8-8-4 cab forward electric locomotive which sold for $11,000. Other trains commanding high bids were a brass Samhongsa 2-

A Mad Mad Mad Scientist laboratory set, one of the showcase items at a Toy Scouts auction, did not receive a single bid until the sale drew to a close — but the item then shot past its presale estimate of $2,500-$4,000 by realizing an auction-leading bid of $6,780.

ABOVE: A Mint in Box Ideal Samantha doll was the top seller at Toy Scouts' annual movie and TV auction, realizing $4,296. ABOVE RIGHT: Even a promotional flyer for the doll brought $220.

8-8-4 locomotive, estimated at $600-$800 but realized $1,292; an O-gauge Union Pacific 4-8-8-4 locomotive at $2,255; and a showman's engine made by Ketleys of Chelmsford at $5,115.

A Buddy-L outdoor railroad set, including a locomotive, four cars, turntable, train shed and track sold for $2,200, at the top of the list for a Richard Opfer auction.

Included in transportation toys were the model ships; another popular item at the 1998 auctions. The Garth Auction featured some prized gems, including a non-motorized model of the Ellen May. The ship, done in 1:85 scale, doubled its presale estimate of $500-$1,000 when it sold for $2,200.

A standard-scale model of the tug Zwarte Zee realized $3,520, while a model of a gas-fired steam clinker received bids of $2,860.

Remote control ships were also present, with a Royal Navy mine hunter Hubberston model making an appearance at $660.

The Bertoia Auction also featured a few model boats. Although small in number, these boats commanded huge

In Search of the Perfect Doll

While many nationwide auction houses conduct general auctions, several specialize in dolls of many eras. When looking for that perfect doll to begin or add to your collection, you might want to contact these specialty companies.

Theriault's
P.O. Box 151
Annapolis, MD 21404
800-638-9422/410-224-3655

McMasters Doll Auctions
P.O. Box 1755
Cambridge, OH 43725
800-842-3526/614-432-4419

Withington, Inc.
RD2 Box 440
Hillsboro, NJ 03244
603-464-3232

prices: an Orkin Craft tin motor boat, sold at $3,025; and a Marklin Rheinland tin battleship, sold at $6,600.

Two lithographed tin and pressed steel rowboats from the 1930s were sold at the Barrett auction. Both were Popeye rowboats; the licensed one sold for $2,310; the unlicensed one sold for $1,210.

Cartoon Extravaganza

Popeye items took the Barrett

No. 93390 **STAR WARS** SIX PACK Ages 4 and up

INCLUDES ALL SIX ACTION FIGURES SHOWN

IG-88

Yoda

Kenner

ABOVE: Collectors in the market for rare, vintage **Star Wars** *toys did not have to look any further than a Contemporary Relics sale. One of the highlights was an unusual six-figure pack that realized $1,000. BELOW: One of the rarest* **Star Wars** *toys ever produced, this Chewbacca hand puppet brought $1,925.*

auction by storm; they were described as "the most diverse offered at the sale." For example, a Popeye Spinach Motorcycle Cart realized $2,530, while painted cast-iron figures of Popeye, Olive Oyl and Wimpy sold for $990.

Many other cartoon characters made their appearance at the auction. Nifty's Felix Sparkler toy sold for $3,190, while a box of 12 Schoenhut Felix Jr. jointed wood dolls sold for $2,970. Chein's painted wood Krazy Kat Express train pull toy realized $1,650, and the 7-inch Krazy Kat velvet doll brought $990.

Betty Boop was a popular girl at the Barrett auction as well. A Betty Boop celluloid nodder with its original box sold for $1,650, while a ukulele featuring images of Betty Boop, Bimbo and Koko sold for $231. A rare, large cloth Betty Boop doll received bids of $104.

Since the word "cartoon" is almost synonymous with Disney, very few cartoon collections exist without some form of Disney characters. The auctions for 1998 were no exception.

Mickey Mouse has always been popular with collectors, and it comes

as no surprise that he commands top dollar at various auctions. A Distler Mickey Mouse Hurdy-Gurdy, one of only six known to exist, sold for an incredible $15,900 at a 'Tiques auction. Even a 1937 Mickey Mouse Air Pilot book brought $161.

Mickey's pals also made an appearance, much to the delight of collectors. Other top Disney sellers included a chalk Donald Duck figure, $1,490; a Christmas store display from the 1930s featuring Donald Duck, Mickey Mouse, Clarabelle, Pluto and Santa

Guide to Major Auction Houses

Want to find out more about some of the nation's leading auction houses or dates of 1999 auctions? Here are some addresses:

• Noel Barrett Antiques & Auctions, Box 300, 6183 Carversville Rd., Carversville, PA 18913 (215-297-5109).
• Bill Bertoia Auctions, 1881 Spring Rd., Vineland, NJ 08361 (609-692-1881).
• Butterfield & Butterfield, 220 San Bruno Ave., San Francisco, CA 94103 (415-861-7500), 7601 Sunset Blvd., Los Angeles, CA 90046 (213-850-7500).
• Christie's, 502 Park Ave., New York, NY 10022 (212-548-1119).
• Contemporary Relics, 1224 Boston Ave., Suite 202, Flint, MI 48503 (810-233-3202).
• Henry/Peirce Auctioneers, 1525 S. Arcadian Dr., New Berlin, WI 53151 (414-797-7933).
• Gene Harris Antique Auction Center, 203 S. 18th Ave., P.O. Box 476, Marshalltown, IA 50158 (800-862-6674).
• Randy Inman Auctions, P.O. box 726, Waterville, ME 04903 (207-872-6900).
• James Julia Inc., P.O. Box 830, Fairfield, ME 04937 (207-453-7125).
• Just Kids Nostalgia, 310 New York Ave., Huntington, NY 11743 (516-423-8449).
• Brian Rachfal auctions, 1177 Branham Lane, Ste. #359, San Jose, CA 95118 (408-629-3980).
• Skinner Inc., 357 Main St., Bolton, MA 01740 (508-779-6241).
• Sotheby's, 1334 York Ave., New York, NY 10021 (212-606-7176).
• Toy Scouts, 137 Casterton Ave., Akron, OH 44303 (330-836-0668).
• 'Tiques, RR1, Box 49B, Rte. 34, Old Bridge, NJ 08857 (732-721-0221).

Claus, $480; an Emerson Snow White and the Seven Dwarfs radio, $3,036; and a Snow White doll, $403.

In the area of figurines, cartoon characters were not all that was represented. Beatles memorabilia also hit high notes with bidders. A rare set of 9-inch figurines sold for $11,500 while a set of 5-inch Remco Beatles dolls realized $880. Even Mego's KISS dolls were represented; a complete set brought in $1,220.

Robots, Space Toys Sold Well

Robot and space toy collectors alike had an incredible year at 1998 auctions. Whereas top selling items tended to lean toward banks and cars, the dolls managed to hold their own, landing well within the higher price ranges.

Some of the highest selling toys at

RIGHT: This spectacular TippCo Mickey and Minnie tin motorcycle wowed the crowds at a Phillips toy and doll auction by bringing $27,000. For more information about Phillips sales, contact Phillips at 7447 Forsyth Blvd., St. Louis, MO 63105, 314-726-5515.

the Henry/Peirce auction were the robots. A wind-up tin 8-1/2-inch Mr. Robot sold for an impressive $1,150, while another robot commanded prices of $2,250. Robby the Robot, one of the most famous mechanical men of all time, realized $1,600; and the Space Commando robot brought $1,400.

In addition to the popular robots, the Henry/Peirce Auction included such items as a Buck Rogers 25th Century laboratory set, $2,600; a tin wind-up Powerful Katrinka, $1,500;

and a battery-operated Mr. Mercury, $425.

Collectors in the market for rare, vintage *Star Wars* toys did not have to look any further than a Contemporary Relics' phone auction.

The highlight of the sale was a rare three-figure set of Princess Leia in Hoth gear, R2-D2 with sensorscope and Luke Skywalker in Hoth gear. Three-figure sets are rare production items that were available in limited supply when issued nearly 20 years ago.

The sets were discovered only recently — one of which was offered to Contemporary Relics by veteran

Jumeaus command top bids at many sales of vintage dolls. A 20-inch Tete Jumeau Bebe sold for $3,575 at a Cobb's auction.

Highlighting a Theriault's auction was a six-piece, original American cloth storybook character set from **Alice in Wonderland.** *The rare storybook dolls sold as a set for $67,000.*

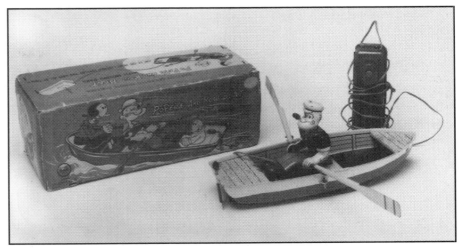

ABOVE: *Popeye tin toys are extremely valuable, as was evidenced at a Shine Gallery sale. A Popeye and a Row Boat sailed away with a price of $8,910. LEFT: Linemar's Popeye Air-O-Plane sold for an eye-popping $19,470 at the sale. For more information on Shine Gallery auctions, contact them at 2934-1/2 N. Beverly Glen Circle, Suite 260, Bel Air, CA 90077, 213-937-7486.*

collector John Wooten. The three-figure pack realized an auction-leading $4,175.

Another rate item — Regal Toy Limited's 3-1/2-foot plush Chewbacca puppet — was also offered. The puppet, made by the Canadian firm, was one of only two or three known to exist. It sold for $1,925.

Another interesting item was a lucite star with the words "May the Force Be With You" engraved on it. Several of the stars were made for George Lucas to be given to cast and crew members. This item realized $650.

Excellent prices were realized for

BELOW: *Foremost among the fine Kewpies sold at a Theriault's auction was this all-bisque figurine of a Kewpie feeding a Doodledog with a baby bottle. It sold for $4,800 compared to a presale estimate of $2,500-$3,000.*

rare non-sports trading cards and boxes and hard-to-find Star Wars toys at a Brian Rachfal auction. A rare Marsh Attacks! trading card display box from 1962 brought $6,098 from a Louisiana physician. The price was a

record for this item.

Collectors engaged in intense bidding for prototype *Star Wars* action figures — the telescoping Darth Vader and Ben Kenobi. Each was a hand-

ABOVE LEFT: *Max won the hearts of many doll lovers at McMasters' Best Friends Sale, but it took $14,500 to bring home this charming character from Kammer & Reinhardt. ABOVE RIGHT: Wendy also drew many admirers at the auction. The 13-1/2-inch Bruno Schmidt character realized $14,000.*

finished pre-production sample from Kenner's original 3-3/4-inch *Star Wars* line. The figures sold for a combined $5,560.

Other *Star Wars* lots included an extremely rare *Return of the Jedi* proof card ($953), a carded version of the die-cast Darth Vader TIE Fighter ($1,575) and a stunning Kenner *Empire Strikes Back* merchandise hanger that closed at $840, about $700 above presale estimates for the item.

Top-Selling Dolls

Although many dolls commanded lower prices, there were a few that topped the list in some auctions. This was the case with Theriault's four-auction weekend, where doll sales topped the charts.

A six-piece original American cloth storybook character set from *Alice in Wonderland* brought in a stunning $67,000, more than twice its presale estimate.

A French bisque first period Bebe known as portrait was the runner-up, commanding $53,000 compared to a presale estimate of $20,000-$30,000. Other bisque dolls included a classic French 14-inch Bebe, $20,000; a Bebe Gigoteur by Jules Steiner, $4,200; and a German bisque child known as the AT-Kestner, $17,000.

Kewpie dolls also made an appearance, with a Kewpie on a toy riding horse, $5,500; Kewpie on a puppy, $5,300; Kewpie on a goat, $5,100; and Kewpie riding a stork, $5,000.

Jumeau dolls were the center of attention at Cobb's Doll Auctions. A 26 inch EJA Bebe #12 in an original dress and shoes brought $22,000; a 28 inch long-faced Triste Bebe #13 with a closed mouth sold for $13,750.

BELOW: A rare Superman premium was the auction's top seller, realizing $9,500

Disneyana brings a lot of interest at many auctions. This silver-plated Mickey Mouse radio was a rare discovery. It paced the Disney lots at a 'Tiques sale by realizing $7,749.

Many fashion dolls were also present at the auction, including a 17-inch Bru Poupee F Smiling Fashion that realized $3,630. Bids of $1,980 bought a 15-inch French Fashion Lady #3 and $1,925 was the winning bid for an 18-inch French Fashion.

German J.D Kestner character babies commanded high prices as well, with a 26-inch #245 Hilda baby realizing $8,800. Two Kestner pouties realized winning bids of $2,475 for a 22-inch child and $1,155 for a 16-inch #VIII child.

Vintage dolls tugged at collectors' heartstrings at a McMasters sale in 1998. The auction's top seller was a German 13-inch Kestner 243 Baby, realizing $4,700. Following close behind it was a 26-inch Jumeau fashion doll at $3,900, just short of its presale estimate.

Bisque dolls are a favorite among collectors, and Theriault's appealed to the masses with its Morning Glories, Evening Joys auction. German and French bisque dolls were in the limelight, with some of the finest signed K.P.M. porcelain dolls ever displayed.

An unusual brown-haired lady sold for $13,500; the Morning Glory Lady sold for $10,250; a lady with a patented Lacmann body sold for $5,100; and a brown-haired gentleman brought $7,000.

An exquisite pair of Swiss artist dolls, hand-made by Sasha Morgenthaler, were the highlight of the auction. The first doll, identified as Type 1, model C, sold for $8,500, while the other, a Type 1, model A, sold for $7,600. Both dolls were gifts from Morgenthaler to the original owner, which helped to increase their value.

A French papier mache automaton known as the Fruit Seller by Gustave Vichy led the automaton category by realizing an auction-leading $21,500. Another automaton was a German Papier mache keywind nodder known as Mr. Rabbit. This automaton sold for $5,600.

The Bertoia auction paid tribute to doll collectors as well, offering a hand-painted tin street musician that sold for $1,045. One of the highest bids of the auction ($18,700), went to

Even television character dolls are turning up at auction. A Mego Toni Tennille from 1977 sold for $45 at McMasters' Shining Star sale, and a Hasbro Charlie's Angels gift set from the same year brought $140.

a rare Lehmann depiction of two musical clowns.

Christmas collectors chose from numerous Santa collectibles also, such as a German hand-painted Santa Claus that sold for $2,420.

Prices were 'Bear'able

Well within the category of dolls is the stuffed teddy bear. Collectors across the country love teddy bears, and bear enthusiasts recognize Steiff bears as leading collectibles. This was evident at the Richard Opfer auction, where the highest bid of $15,000 went to a honey-colored mohair Steiff bear. The fully-jointed teddy bear featured an underscore button, shoe-button eyes and a non-working growler.

McMasters auction offered a 37-inch cinnamon plush bear that realized $225 at their doll auction.

Bears worked their usual magic on eager bidders at Cobb's. A bid of $2,695 took home a 16-inch Steiff hump back with honey-colored mohair.

RIGHT: Mechanical and still banks are popular sellers at many auctions. This Billie Jean King and Bobby Riggs sold for $425 at a Henry/Peirce sale. BELOW LEFT: Speaking Dog was the auction's top seller, realizing a whopping $7,900.

Other unusual bears included a 14-inch unmarked golden mohair bear ($800), and a 16-inch dark auburn mohair bear with a spring-hinged mouth brought $550. The bear's mouth opened to reveal a throat and chest cavity.

Bent on Barbie

Barbie is Mattel's star doll, and collectors have been in love with her since she first appeared in 1959. The year 1998 could once again be considered the year of Barbie, since collectors just can't seem to get enough.

Barbie made an appearance in many auctions, sometimes dominating

the sales. Such was the case with the properly dubbed Shining Star McMasters auction. The two top sellers of the auction consisted of a set and a doll.

The set, including a Ponytail Barbie #3 with resort and picnic apparel, sold for $1,400, while the highest bid doll, a brunette Barbie from 1970 with a ribbon bow and Living Barbie and Skipper booklet, realized $1,400.

American Girl Barbies were also popular with bidders, as shown by an American Girl that sold for $800 even though one leg was slightly longer than the other.

Other popular Barbies at the auction included a Bubblecut Barbie from 1961 ($95), and another doll from the same period with titian hair ($80).

Barbie's friends also made an appearance and brought some respectable bids. A Living Skipper from 1970 sold for $55, a blond Ken from 1965 realized $100, and a blonde Midge from 1964 brought $70.

A Doll Express auction, devoted entirely to Barbie and Barbie accessories, kept 218 attendees occupied with more than 600 lots. The top seller, an American Girl Side-part, sold for a whopping $3,600. Down from there but still one of the top sellers was a Ponytail Barbie #4 that sold for $250.

Twist and Turn Barbie and Barbie's friends were also popular items, selling for $175-$375 apiece.

The first Bob Mackie Barbie, called a Bob Mackie Designer Gold, from 1990 brought $400. The Happy Holidays series also made a modest

Cast iron and pressed steel vehicles often bring good prices at auction. A Silver Flash Racer sold for $2,860 at Bill Bertoia Auction's Summer Toy, Train & Doorstop sale. This interesting Chein piece included a driver, exhaust pipes, wind deflector, a spare tire and a grille that read "Silver Flash #8."

While antique dolls are the centerpiece of many doll auctions, vintage (those from the 1960s-1970s) Barbies and even newer Barbies are also turning up at more and more auctions. This swirl Ponytail Barbie from 1964 brought $400 at McMasters' Shining Star sale.

appearance, with dolls from 1992 selling for $160 and a doll from 1994 selling for $150. These dolls originally cost only $30 retail new.

Minnesota also devoted an entire auction entirely to Barbie, and collectors weren't disappointed with the 591-lot selection. The top seller, a Ponytail Barbie #3 blonde from 1961, realized $1,045.

All Dolled Up

Enough people collect vintage doll clothing to warrant an auction, so Theriault's believed in 1998. The auction, titled What Dolls Wore Before, treated attendees to an incredible collection of vintage (1850-1925) doll clothing.

The sale included 450 lots of antique doll collections, including a surprising price on a Russian wedding costume for an early French Bebe, of $11,400. The costume was only estimated at $2,000-$2,500.

Also making an appearance in the auction was an ivory satin two-piece gown in its original box, earning an eye-popping $27,500.

Odds and Ends

Out of all the standard collectibles, there are always a few unusual items that aren't numerous enough to warrant their own category. This does not make them any less popular, however, as auctioneers have found out in 1998.

These items range from television

dolls to board games to *TV Guide* magazines. The Toy Scouts' annual movie and TV auction was a superb example.

The auction's top seller, a Samantha doll from the television series *Bewitched*, realized $4,296. This far exceeded its presale estimate of $2,500-$3,000. Even a promotional flyer for the doll managed to top out at $220.

Some of the more outstanding prices were commanded by *TV Guide* magazines. These far exceeded their presale estimates, including a 1965 edition of *TV Showtime* ($586) featuring *Lost in Space* stars Guy Williams and June Lockhart. It was estimated at only $50-$75!

An issue of *TV Showtime* from 1964 featuring *The Munsters* fared even better, earning $700 compared to the presale estimate of $35-$60.

The top seller, a Mad Mad Mad Scientist Laboratory Set, was also the item that remained idle the longest. The set did not receive a single bid until the closing of the show, where it suddenly shot past its presale estimate of $2,500-$4,000 and topped out at $6,780.

Superman secret candy compartment rings are the Holy Grail of premium collecting. Leader Novelty issued these now rare Superman rings for an incredibly limited period in the early 1940s. No wonder these premiums bring such super prices today. That included a recent 'Tiques auction in which a Superman candy ring brought an auction-leading $9,500.

A silver-plated metal Mickey Mouse radio, featuring a raised Mickey beating a drum, more than doubled its presale estimate by realizing $7,749 at a 'Tiques sale.

Other unique Disney items included a Mickey night light holder in its original box that sold for $2,890 and a 6-piece porcelain Mickey orchestra that brought $3,970.

Showcases of Disney collectibles, trains, tin and pressed steel vehicles, mechanical banks and even doorstops were crammed to capacity at Bertoia's Summer Toy, Train & Doorstop Sale. Anchoring the sale's highlights was a fabulous grouping of hand-painted doorstops. The selection was the best collection offered at market for many years. Collectors battled for the prized Tavern, which finally sold for $5,280. Other headliners included Vase of Flowers ($2,750), White Cockatoo ($1,760) and a Hubley Popeye ($2,750).

The sale also featured Disney toys from the famed Ernest Trova collection. They included such celluloid

Doorstops were showstoppers at a Bertoia sale. They included, from left — a Hubley Popeye, $2,750; Rhumba Dancer, $2,200; Police Boy, $1,100 and Donald Duck, $550.

desirables as the Donald on Rocker ($4,070) and rare Mickey prototype roller skater that brought $5,500.

But it was a turn-of-the-century mechanical bank that brought the auction's top price. J&E Stevens' Calamity realized $16,500, topping the bank's presale estimate of $8,000-$10,000. The gridiron piece features two tackles and a fullback.

MAD About Collectibles

The sixth annual MAD collectibles auction was once again a complete success. Both participants and auction coordinators were happy with the results.

"With long-term and new fans of *MAD* magazine on the prowl for those elusive MAD collectibles, we were excited with the incredible variety of MAD material that we had up for bid," said auction coordinator Michael Lerner.

The top selling item was a 1960 Alfred E. Neuman Mardi Gras costume, tripling its presale estimate

LEFT: Japanese and other colorful battery-operated robots are becoming increasingly popular with collectors. Answer Game Machine Robot brought $550 at Bertoia's Summer Toy, Train & Doorstop Sale.

BELOW: Collegeville's Alfred E. Neuman Mardi Gras costume sold for $1,611 — more than three times its presale estimate.

with a high bid of $1,611.

One of the biggest surprises of the auction was the sale of business cards from late MAD publisher William M. Gaines, which sold for $150.

Other auction highlights included a MAD Style Book ($150), set of 1950s British *MAD* magazines ($220), Alfred E. Neuman bust ($450) and Aurora's Alfred E. Neuman model kit ($250).

Market Update

A Look at 1998's Secondary Market for Toys

By Guy-Michael Grande

What toys have been hot — or not — on the secondary market for the past 12 months? It may surprise you to see just how well, or how poorly, some of your favorites have done.

This article surveys the secondary market action in several areas this past year. Also featured are lists of the top-selling toys in various catego-

Action figure exclusives — like this Wal-Mart exclusive **Star Wars** *Cantina Band member — are holding steady as collector favorites.*

ries. Prices are gathered from a variety of sources and are featured in the Krause Publications' title, *1999 Toys & Prices*.

This market update gives you a sneak peek at *1999 Toys & Prices*. With more than 900 pages, the book includes updated pricing and expanded listings for more than a dozen categories of toys ranging from action figures to vehicle toys.

Produced by the publishers of *Toy Shop*, the book is available for $18.95 plus shipping and handling from Krause Publications, 800-258-0929.

Action Figures

Whether scaled to fit in the palm of your hand or stand up to a foot tall, action figures remained a commanding presence in the wild world of toy collecting. Heroes and heroines mingled with monsters, robots and villains, even sharing shelf space last year with race car drivers, rock stars and revamped classic characters.

Released in conjunction with George Lucas's retooled *Star Wars* trilogy in 1997, Kenner's *Star Wars* figures continued their dominance of retail shelves this year. While interest was fueled by early demand for short-carded figures and exclusives, some secondary market insiders cite the sheer number of characters now being manufactured as both a boon and bane for collectors.

Never has there been such an extensive, detailed line as Kenner's current *Star Wars* offerings, which has caused some frustration for completists and an apparent glut of

POWER OF THE FORCE — **Kenner's newer Star Wars** *figures, like the Luke Skywalker / Wampa set pictured, sold well for action figures dealers in 1998.*

figures at the retail level. Kenner's 12-inch characters — especially those offered originally as store exclusives — captured many a collector's imagination, including the relatively scarce AT-AT Driver.

Disappointing as *Batman & Robin* the movie may have been, secondary market

The Top 10 Mechanical Banks
(in Excellent condition)

1. Mikado, Kyser & Rex, 1886 . . . .$55,000
2. Jonah and the Whale, Jonah Emerges, Stevens, J.& E., 1880s . . .55,000
3. Girl Skipping Rope, Stevens, J.& E., 1890 45,000
4. Circus Bank, Shepard Hardware, 1888 45,000
5. Calamity, Stevens, J.& E., 1905 35,000
6. Rollerskating Bank, Kyser & Rex, 1880s 30,000
7. Harlequin, Stevens, J.& E., 1907 22,000
8. Turtle Bank, Kilgore, 1920s 20,000
9. Picture Gallery Bank, Shepard Hardware, 1885 20,000
10. Motor Bank, Kyser & Rex, 1889 20,000

The Top 10 Still Banks
(in Excellent condition)

1. Indiana Paddle Wheeler, Unknown manufacturer, 1896 . . . .$8,000
2. Tug Boat, Unknown manufacturer 7,500
3. San Gabriel Mission, Unknown manufacturer 7,500
4. Coin Registering Bank, Kyser & Rex, 1890 6,500
5. Hippo Bank, Unknown manufacturer .6,000
6. Dormer Bank, Unknown manufacturer 6,000
7. Chanticleer (Rooster), Unknown manufacturer, 1911 6,000
8. Boston State House, Smith & Egge, 1800s 6,000
9. Electric Railroad, Shimer Toy, 1893 .6,000
10. Camera Bank, Wrightsville Hardware, 1888 5,000

The Mikado cast-iron mechanical bank — when found at auction — can top $25,000 easily in prime condition.

demand for Kenner's four 12-inch figures remained strong, especially for Batgirl — only the second 12-inch version of her character created since Ideal's legendary Super Queen (1967).

Demand for Kenner's original Batman: The Animated Series line escalated further, as did overseas demand for figures and vehicles from the first two Batman films, raising the Dark Knight's values to new heights.

Todd McFarlane continued to rock the toy industry in his inimitable fashion,

launching his second lineup of KISS figures based on McFarlane's Psycho Circus comic book.

McFarlane displayed his company's keen competitiveness by capturing the license for *The X-Files* figures as the popular Fox TV series made its transition to the silver screen. One glance at the renditions of Mulder and Scully confirmed McFarlane's unrivalled excellence in both character likeness and detail.

Hasbro's G.I. Joe seemed to commandeer a

Action figures by McFarlane Toys fare consistently well on the secondary market. Pictured are toys from McFarlane's hot 1998 line, KISS Psycho Circus.

Whether you love them or hate them, Beanie Babies continued their reign in 1998. Long retired Beanies continue to command the most; recent issues are slipping in value.

The Top 10 Beanie Babies
(All listed are retired.
Prices are for items in Mint condition)

1. Peanut the Elephant, dark blue $4,200
2. Bongo the Monkey, with Nana tag 4,000
3. Punchers the Lobster 4,000
4. Derby the Horse, old tag, fine yarn 3,800
5. Cubbie [Brownie] the Bear, old
   "Brownie" tag 3,700
6. Teddy the Bear, old face 2,500
7. Quackers the Duck, without wings 2,000
8. Humphrey the Camel 1,800
9. Slither the Snake 1,750
10. Teddy the Cranberry Bear, old face 1,600

The Top 10 Barbie Dolls
(in Mint in Box condition)

1. Ponytail Barbie #1, brunette, 1959 . . . . . .$9,000
2. Ponytail Barbie #1, blonde, 1959 8,000
3. Ponytail Barbie #2, brunette, 1959 7,000
4. Ponytail Barbie #2, blonde, 1959 6,000
5. American Girl Side-Part Barbie, 1965 4,000
6. Color Magic Barbie, midnight black
   hair, 1966 3,200
7. American Girl Barbie, brunette,
   titian, 1966 2,700
8. American Girl Barbie Color Magic Face,
   1966 2,700
9. Color Magic Barbie, blonde,
   plastic box, 1966 2,400
10. Pink Jubilee Barbie, only 1,200 made, 1989 2,200

The Top 10 Barbie Fashions
(in Never-Removed-From-Box condition)

1. Roman Holiday, #968 . . . .$4,800
2. Pan American Stewardess, #1678 4,500
3. Easter Parade, #971 4,000
4. Gay Parisienne, #964 3,800
5. Beautiful Bride, #1698 2,100
6. Shimmering Magic, #1664 1,800
7. Here Comes The Groom, #1426 1,600
8. Campus Sweetheart, #1616 1,500
9. Commuter Set, #916 1,500
10. Gold n' Glamour, #1647 1,500

The Top 10 Toy Guns / Sets
(in Mint in Box condition)

1. Man From U.N.C.L.E. THRUSH Rifle, Ideal, 1966 . . . .$2,500
2. Lost In Space Roto-Jet Gun Set, Mattel, 1966 2,000
3. Man From U.N.C.L.E. Attache Case, Ideal, 1965 1,500
4. Cap Gun Store Display, Nichols, 1950s 950
5. Showdown Set with Three Shootin' Shell Guns, Mattel, 1958 950
6. Man From U.N.C.L.E. Attache Case, Lone Star, 1966 850
7. Man From U.N.C.L.E. Napoleon Solo Gun Set, Ideal, 1965 800
8. Lost In Space Helmet and Gun Set, Remco, 1967 800
9. Roy Rogers Double Gun & Holster Set, Classy, 1950s 800
10. Man From U.N.C.L.E. Attache Case, Lone Star, 1966 750

She may be used to fancy dresses, but Mattel's Barbie had lots of careers in 1998. Last year, she even joined the NBA — look out, Michael Jordan!

smaller stretch of the toy battlefield in 1998, if he didn't lose any great degree of ground among collectors. As several 3-3/4-inch figures were being reissued in three-packs at retail, secondary market demand for the Female Helicopter Pilot — only the second 12-inch female figure in G.I. Joe's decorated history — remained airborne. Of the smaller vintage 3-3/4-inch Joe figures, dealer Perry Mohney of Maryland-based Toy Exchange says

"It's just phenomenal how many of those I sell."

In his region, Mohney admitted demand for carded vintage *Star Wars* figures had "really died off, and what I've found out is the people that were collecting them are trying to keep up with the new stuff, before it goes up. They're putting their money into the new stuff," he summarized. "To me, the old stuff's on the back burner right now; the prices have stabilized. I'm still selling [it], but not like I used to."

Mohney did confirm brisk sales and continued collector interest in new *Star Wars* and the original Spawn figure line, as well as "any of the female action figures," regardless of their manufacturer, singling out McFarlane Toys' limited release Red Angela.

"Transformers are hot, too," he added. "They're coming in to their time. I do really well with them." As for future trends, he pointed to the often overlooked He-Man and Thundercats figure lines, as the population of children who grew up with these toys is coming into their 20s — prime time for recapturing their childhood treasures — and their values are on the rise.

Banks

Prewar cast-iron mechanical banks still commanded the highest interest rates among collectors, as last year's market demonstrated. A Speaking Dog yipped its way to $7,900 when auctioned mid-year by Henry/Pierce Auctioneers. The classic Shepard Hardware bank was closely followed by Ole Storle's Butting Ram ($5,600) and a bronze Hippo still bank ($5,400) despite its unknown origin.

Auction highlights also included the Kress Building ($3,400), Kyser & Rex's 1899 Motor Bank ($3,300), Shepard Hardware's

The Top 10 Character Collectibles
(in Mint condition)

1. *Action Comics #1*, DC Comics, 1938, first appearance of Superman $100,000
2. Superman Member Ring, 1940 100,000
3. Superman Gum Ring, Gum, Inc., 1940 40,000
4. Superman Candy Ring, Leader Novelty Candy, 1940 22,000
5. Superman Trading Cards, Gum Inc., 1940, set of 72 10,000
6. Superman Patch, 1939, Supermen of America, Action Comics 10,000
7. Superman-Tim Club Ring, 1940s 10,000
8. Captain Marvel Sirocco Figurine, Fawcett, 1940s 8,500
9. Donald Duck Bicycle, Shelby, 1949 .6,000
10. Superman Patch, 1940s 5,500

Jonah and the Whale ($3,200), Reynolds Foundry ($3,100), J&E Stevens' Hen and Chick ($3,000), Indian Shooting Bear ($2,400) and Kyser & Rex's Globe Savings Fund Bank from 1899 ($2,300).

Still gaining popularity among newer collectors, die-cast banks were once more the benefactors of increasing interest. Top manufacturer Ertl continued its dominance over this field with its superbly manufactured limited edition pieces. Industry insiders expect this segment of the market to demonstrate steady growth among new collectors who are eager to deposit their toy dollars into these attractive yet somewhat more affordable investments.

Barbie

Limited-edition Barbies continued to share the spotlight with vintage issues among Barbie collectors last year. The leather-clad Harley-David-son Barbies (two versions were made), NASCAR Barbie and the Barbie and Ken X-Files gift set, were sought after both at retail and as secondary market pieces.

Thanks to an unfortunate packaging recall that some say was due to reasons as mysterious as the storylines which fuel the hot X-Files franchise, the original X-Files set has rocketed up to $175 for those fortunate enough to find one. Sightings of these sets on the secondary market, however, were as rare as confirmed extraterrestrial contacts. Prices for X-Files sets, however, soon dropped once the packages were reintroduced into stores weeks later.

Never on the verge of going out of style anytime soon, vintage Barbie dolls and accessories kept collectors fascinated, enjoying continued demand. "Vintage is always hot," admitted leading Barbie dealer Marl B. Davidson of Florida-based Marl & B. "Everybody wants it because their value goes up all the time."

Next year looks to be even more

The Top 10 Lunch Boxes
(in Near Mint condition)

1. 240 Robert, steel, Aladdin, 1978 ....$2,500
2. Toppie Elephant, steel, American Thermos, 1957 1,600
3. Home Town Airport Dome, steel, King Seeley Thermos, 1960 .1,000
4. Underdog, steel, Okay Industries, 1974 900
5. Knight in Armor, steel, Universal, 1959 825
6. Ballerina, vinyl, Universal, 1960s 800
7. Superman, steel, Universal, 1954 800
8. Dudley Do-Right, steel, Universal, 1962 800
9. Bullwinkle & Rocky, steel, Universal, 1962 800
10. Little Friends, vinyl, Aladdin, 1982 760

The Top 10 Prewar Games
(in Excellent condition)

1. Bulls and Bears, McLoughlin Bros., 1896 ....$13,000
2. New Parlor Game of Baseball, Sumner, 1896 10,000
3. Champion Baseball Game, Schultz, 1889 6,800
4. Little Fireman Game, McLoughlin Bros., 1897 6,000
5. Zimmer Baseball Game, McLoughlin Bros., 1885 6,000
6. Teddy's Ride from Oyster Bay to Albany, Jesse Crandall, 1899 5,500
7. Golf, Schoenhut, 1900 5,000
8. Egerton R. Williams Popular Indoor Baseball Game, Hatch, 1886 5,000
9. Great Mails Baseball Game, Walter Mails Baseball Game, 1919 4,100
10. Darrow Monopoly, Charles Darrow, 1934 4,000

The Top 10 Postwar Games
(in Mint condition)

1. Elvis Presley Game, Teen Age Games, 1957 ....$1,000
2. Red Barber's Big League Baseball Game, G & R Anthony, 1950s 900
3. Win A Card Trading Card Game, Milton Bradley, 1965 ...900
4. Munsters Drag Race Game, Hasbro, 1965 800
5. Munsters Masquerade Game, Hasbro, 1965 800
6. Creature From The Black Lagoon, Hasbro, 1963 750
7. Strike Three, Tone Products, 1948 725
8. Jonny Quest Game, Transogram, 1964 700
9. Munsters Picnic Game, Hasbro, 1965 700
10. Gilligan's Island, Game Gems, 1965 600

The Top 10 PEZ
Dispensers
(in Mint condition)

1. Make-A-Face, American card $3,300
2. Elephant 3,000
3. Witch Regular 3,000
4. Make-A-Face, German "Super Spiel" card 3,000
5. Mueslix, European 2,900
6. Pineapple 2,750
7. Lion's Club Lion 2,500
8. Make-A-Face, no card ..2,200
9. Space Trooper, gold ...1,900
10. Bride 1,900

fashionable for Mattel's top model, as the toy giant will undoubtedly pull out all the stops when celebrating Barbie's 40th anniversary.

Beanie Babies

Within a relatively short period of time, Beanie Babies have become the most wanted toys to hit the collectibles market in years. You can buy them at retail stores and on the secondary market, get them through special restaurant premium promotions and at major league ballparks, or order them via cable TV shopping channels, through magazines and over the Internet. Simply put, Beanie Babies are everywhere. Love 'em or hate 'em, they are a phenomenon that

FORGET THE FOOD! — The most-anticipated fast food event of 1998 was McDonald's release of Teenie Beanie Babies.

won't be ignored.

Retired characters — along with those believed to be harder to find because of lower production numbers or regional availability — have so far shown the greatest appreciation in value. A seemingly limitless number of manufacturers are trying to ride the tidal wave of Beanie Baby popularity which Ty, Inc. started and continues to promote with new issues and retirements.

McDonald's Teenie Beanie success reportedly had the toys selling better than 99-cent Big Macs, and secondary market demand continued to be strong for these issues, especially Bongo the Monkey.

Last year also saw

Disney jump into the fray as it launched its own line of beanbag toys — using their own iconic characters, of course — with Warner Brothers similarly following suit. Though debatable whether or not Beanie Babies will remain white-hot collectibles, their increased popularity last year presented a veritable bonanza for Beanie enthusiasts.

The Top 10 Space / Science Fiction Toys (in Mint in Package condition)

1. Lost In Space Doll Set, Marusan/Japanese $7,000
2. Buck Rogers Solar Scouts Patch, Cream of Wheat, 1936 5,500
3. Buck Rogers Cut-Out Adventure Book, 1933 4,500
4. Space Patrol Monorail Set, Toys of Tomorrow, 1950s 4,000
5. Buck Rogers Roller Skates, Marx, 1935 3,500
6. Space Patrol Lunar Fleet Base, 1950s 2,600
7. Buck Rogers Pocket Watch, E. Ingraham, 1935 2,500
8. Lost In Space Switch-and-Go Set, Mattel, 1966 2,300
9. Buck Rogers Costume, Sackman Bros., 1934 2,100
10. Lost In Space Roto-Jet Gun Set, Mattel, 1966 2,000

The Top 10 Fast Food Toys (in Mint condition)

1. Big Boy Nodder, Big Boy, 1960s $1,500
2. Black History Coloring Books, McDonald's, 1988 400
3. Big Boy Bank, large, Big Boy, 1960s 300
4. Big Boy Bank, Medium, Big Boy, 1960s 165
5. Colonel Sanders Nodder, Kentucky Fried Chicken, 1960s 150
6. Transformers, McDonald's, 1985 140
7. Big Boy Board Game, Big Boy, 1960s 120
8. Big Boy Bank, small, Big Boy, 1960s 100
9. Triple Play Funmeal Boxes, Burger Chef, 1977 90
10. My Little Pony, regional promo, McDonald's, 1985 80

The Top 20 Tin Toys (in Mint condition)

1. Popeye the Heavy Hitter, Chein . .$6,500
2. Popeye Acrobat, Marx 5,500
3. Popeye the Champ, Marx 4,600
4. Mikado Family, Lehmann 3,900
5. Red the Iceman, Marx 3,500
6. Popeye with Punching Bag, Chein .2,500
7. Lehmann's Autobus, Lehmann . . .2,500
8. Mortimer Snerd Hometown Band, Marx 2,500
9. Popeye Express, Marx 2,400
10. Hott and Trott, Unique Art 2,300
11. Superman Holding Airplane, Marx 2,250
12. Mortimer Snerd Bass Drummer, Marx 2,200
13. Tut-Tut Car, Lehmann 2,200
14. Ajax Acrobat, Lehmann 2,200
15. Hey Hey the Chicken Snatcher, Marx 2,200
16. Zig-Zag, Lehmann 2,000
17. Popeye Handcar, Marx 2,000
18. Howdy Doody & Buffalo Bob at Piano, Unique Art 2,000
19. KADI, Lehmann 2,000
20. Merrymakers Band, Marx 1,900

Character Toys

Unlike fad toys, which can benefit from stellar demand one day and see the bottom fall out due to inflated values the next, character toys remained a sure bet among collectors last year. Perhaps hundreds of thousands collectibles have featured the most memorable characters from cartoons, comics, TV and film since the dawn of toys, but as was the case last year, premiums remained the single most desirable kind of item in this category.

Godzilla items stomped their way to a comeback of sorts despite a disappointing box office showing for the much-hyped film. Interest in 1960s vintage character toys remained high as more and more properties were tapped for new films like *Lost in Space* and *The Avengers*.

Disney collectibles maintained their commanding presence among character toys last year, joined by vintage superheroes and heroes of the old West. Mickey Mouse, Batman, Superman, Hopalong Cassidy, the Lone Ranger and Roy Rogers continued to enthrall collectors young and old alike. Bugs Bunny, Garfield, the Peanuts gang and Popeye toys still fascinated and illuminated the secondary market.

The 1998 deaths of Roy Rogers, Buffalo Bob Smith and Shari Lewis may have resulted in a slight demand, and increased prices, for

The lunch box hobby may have cooled a little in the past few years, but you can still find plenty of the vintage steel and vinyl classics at toy shows. Among those that fare the best are steel boxes with popular and memorable TV show characters like those pictured above.

related character collectibles.

Fast Food Toys

Competition between fast food restaurant giants continued to be a boon for collectors in 1998. Never have there been more licensed and unlicensed premiums up for grabs in this hot segment of the collectible toys market.

Again generating collector frenzy and media attention, Ty's Beanie Babies hit McDonald's and the secondary market in full force last year. With the Disney license exclusively secured, McDonald's issued a slew of toys based on films like *Mulan*, but their promotions fell short of igniting collectors' imaginations as fully as the giant of silver screen animation had hoped.

"This has been a real strange year," admitted Florida-based fast food toy collector/dealer Pat Sentell. "McDonald's is not as hot as it was a few years ago, I'm saying in general." With the exceptions of the 101 Dalmatians and Teenie Beanie promotions, Sentell feels McDonald's was

The Top 10 Marx Play Sets
(in Mint in Box condition)

1. Johnny Ringo Western Frontier Set, 1959, #4784 . . . .$2,500
2. Civil War Centennial, 1961, #5929 2,000
3. Fire House, #4820 2,000
4. Gunsmoke Dodge City, 1960, #4268 2,000
5. Johnny Tremain Revolutionary War, 1957, #3402 2,000
6. Ben Hur, #4701 1,800
7. Sears Store, 1961,#5490 1,800
8. Custer's Last Stand, 1963 #4670 1,800
9. Wagon Train, #4888 1,800
10. World War II European Theatre, #5949 1,500

The Top 10 Figural Model Kits
(in Mint in Box condition)

1. Godzilla's Go-Cart, Aurora, 1966 . . . .$3,200
2. Lost In Space, large kit w/chariot, Aurora, 1966 1,350
3. Frankenstein, Gigantic 1:5 scale, Aurora, 1964 1,300
4. Munsters Living Room, Aurora, 1964 1,300
5. King Kong's Thronester, Aurora, 1966 1,250
6. Lost In Space, small kit, Aurora, 1966 920
7. Lost In Space, The Robot, Aurora, 1968 800
8. Addams Family Haunted House, Aurora, 1964 800
9. Bride of Frankenstein, Aurora, 1965 750
10. Penguin (Batman), Aurora, 1967 525

eclipsed by, among others, Burger King's Men in Black offerings.

"Beanie Babies were real hot in 1997, of course; that was their first year. And then in 1998, right before the new ones came out, people were realizing they didn't get all of them from '97, and those were real hot and the prices went up. Some of those were bringing $25-$30 apiece for those little ones. Beanie Babies as a whole were way up, but [the general feeling is that] Beanie Babies are getting soft on their prices, too."

Sentell added that Wendy's "has come out with some good toys," citing their Snoopy as one

ABOVE: Toy guns can be found at almost any toy show. BELOW: Who doesn't like PEZ dispensers? Collectors love them; kids love them. Vintage PEZ can be costly, but you can easily start an inexpensive collection of current dispensers. The early Monsters pictured below featured soft rubber heads.

The popularity of vintage Aurora model kits has led Playing Mantis to make its own reissues of these figural favorites under the Polar Lights name. Pictured below is the King Kong's Thronester.

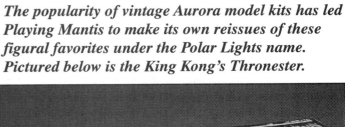

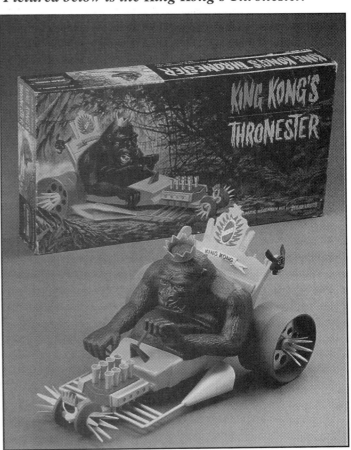

The Top 10 Vehicle Toys (in Mint condition)

1. Packard Straight 8, Hubley $15,000
2. White Dump Truck, Arcade 15,000
3. Checker Cab, Arcade 15,000
4. White Moving Van, Arcade 13,500
5. Elgin Street Sweeper, Hubley 11,500
6. World's Greatest Circus Truck, Keystone, 1930s 10,000
7. Ingersoll-Rand Compressor, Hubley . . .10,000
8. Motorized Sidecar Motorcycle, Hubley 10,000
9. Yellow Cab, Arcade 10,000
10. Ahrens-Fox Fire Engine, Hubley 8,000

The Top 10 Mattel Hot Wheels (in Mint in Package condition)

1. Volkswagen Beach Bomb, surf boards in rear window, 1969 . . . .$4,500
2. Custom Camaro, 1968, white enamel . .2,000
3. Snake, 1973, white/yellow 1,500
4. Mongoose, 1973, red/blue 1,400
5. Mustang Stocker, 1975, white 1,200
6. Carabo, 1974, yellow 1,200
7. Custom Mustang, 1968 1,200
8. Mercedes C-111, 1973 1,200
9. Superfine Turbine, 1973 1,100
10. Ferrari 312P, 1973 1,100

that nobody else has done. She cited Subway and Long John Silver's as two restaurant chains that were rising in the fast food premium race.

"What we're finding is people are going more to toys that they remember, or cartoons or something that they collect, like M&Ms or Jetsons." It seems character-related toys led the market for these premiums last year. Sentell noted that Hardee's Woody Woodpecker "is still a sleeper" because "nobody's ever come out with a Woody Woodpecker. But we all can remember Woody Woodpecker, and that's a neat one."

"There's still new people coming on all the time," she enthused. But while the newer fans seem to seek out the bargains, hardcore collectors are willing to pay higher prices for items like Happy Meal boxes. She added that fewer fast food toys are readily available at tag sales and flea markets "because there are more collectors."

Games

Continuing the trends as mentioned in our previous section, board games based on popular characters retained their hold over the hearts and dollars of eager collectors last year. As it did two years ago, the market for sports-related games once more posted a solid average for fans of legendary

players. Baseball-themed games continued to drive in more home runs than any other sport in this category.

Keeping fresh with current collectors, more regional editions of Monopoly passed "Go" in conjunction with USAOPOLY last year. Among them was the Rhode Island Special Edition, which was designed to benefit the Hasbro Children's Hospital in Pawtucket, R.I. Special editions also commemorated the 50th anniversary of NASCAR and the 95th anniversary of Harley-Davidson.

Toy Guns

Character-related toy guns continued to out-draw

It's tough to find 1960s Marx play sets still in complete condition in original boxes. That's why Mint examples tend to sell for high prices. The current Marx company is making new versions of the old standards.

other shooters in this collectible arena purely on their inherent cross-collectibility. Western-themed sets shared the range once more with their counterparts from detective- and science-fiction based issues. This segment of the secondary market has continued to be the province of older collectors, according to toy dealer Perry Mohney, owner of the Toy Exchange. This gap is perhaps due to the fact that fewer toy guns are currently being made for today's children, and those which are being made are clearly identified

as toys, cast either in bright-colored plastics or tipped with neon caps to differentiate them from the real thing. Mohney says that while infrequent, the toy guns which do come up for sale at his walk-in store move pretty fast.

Lunch Boxes

Made of metal, plastic or vinyl, character-related or otherwise, lunch boxes of all kinds retained their strong hold on collectors in 1998. Leading dealer Larry Aikins of Athens, Texas, shared his thoughts on the past year of lunch box collectibles.

"I list around 60 to 100 lunchboxes a week [as auction sales]," said Aikins, who no longer travels to toy shows.

"A third of them will sell every auction, and we're shipping them literally all over the world. You can get people bidding on your stuff from around the world, and it's really kept us a-jumpin'," admitted the veteran dealer in his unmistakable Texas drawl.

"I bought two collections this last year and

Older pre-1950s board games are a great treat for the game collector. Like the Parker Brothers examples pictured, the games often had interesting themes and graphically-pleasing covers. Prewar games are often hard to track down in Good condition.

am about out of them already. I have a large surplus of lunch kits that I dig into all the time, but it's a-boomin, it's keeping us busy. It's been super."

„But it's a whole new market of people," he ceded. "A lot of them are buying back their old lunch box."

Aikins sold a Holly Hobbie lunch box last year for $65, a high price for such a common box that was in average condition, no less.

"Which is just unheard of," he stated. "Your common boxes are selling, because people can't find them anymore. If you go to the shows or the flea markets, you just don't see any lunch kits. They've dried up. In the last two years is when they really started getting hard to find, but now, they're just next to impossible to find."

"A lot of new people [are] comin' on. I get a call dang near daily from a new collector just starting up," he revealed. "You know, they're talking about how excited they are that they've been looking for three weeks to find a lunch kit."

Aikins said character-related boxes are "doing okay," while "vinyl slowed up for a while, and they're starting to come back a little bit now." He cited the recent reproductions of metal

THE KING IS GONE — Roy Rogers' death in 1998 may lead to increased values for collectibles bearing his name and image. A plethora of Western collectibles exist. Gene Autry's death in 1998 also may generate more interest in other items from the old West characters we knew and loved.

lunch boxes as surprising to him, although the quality was not as good as the classic boxes on which they were based, noting the Howdy Doody repro as having a fuzzy picture and being made of a cheaper gold-colored metal. As for the repros' effect on classic values, "it won't hurt it a doggone bit," offered Aikins.

Marx Play Sets

Marx play sets continued to hold their legendary place among collectors. Prices for the top Mint in Box examples maintained their stability from last year. Aside from their late 1950s-early 1960s vintage, part of their ongoing appeal lies in cross-collectibility.

Western- and war-themed Marx play sets remain the most sought-after, with Johnny Ringo ($2,500) and Johnny Tremain ($2,000) still drawing the most collector demand. Character and Disney-themed issues remain highlydesirable as well, including Jungle Jim ($1,400), Zorro ($1,300) and Ben Hur ($1,250).

"When I get them, they sell," said Toy Exchange owner Perry Mohney.

Model Kits

Kits in 1:18 and other scales from AMT/Ertl, Fujimi, Lindberg, Playing Mantis, Revell-Monogram and Solido kept up the fast pace set a year earlier.

Of the newer issues, garage kits

ABOVE LEFT: Among the most valuable and desirable space toys are Japanese robots like the 1950s tin examples pictured. LEFT: The Amos n' Andy Jalopy is one vintage tin character toy sought by collectors of both tin and of vintage Black Americana. The key to collectible tin is to find items in their best condition because finding tin lithography in top shape is a challenge.

and newer Japanese kits continued to expand their bond on creative model kit builders' imaginations. Among classic model kits, ever-elusive, sealed MIB examples of Aurora kits still outmodeled everything else on the secondary market. Godzilla's Go-Cart ($3,000), the Munsters Living Room ($1,200) and King Kong's Thronester ($1,000) dominated collector demand.

Playing Mantis's reissues of vintage Aurora kits (under the Polar Lights name) generated much buzz in the collector market.

PEZ

Affordable, compact and available nearly everywhere you go, an increasing number of collectors have been giving in to their sweet tooth when it comes PEZ dispensers. While the older candy dispensers still command prices for those with expensive tastes — ranging from $500 to as high as $3,000 each — locating examples at both the retail and secondary market levels remains easy for even the average PEZ collector.

The Make-A-Face versions, which are similar to a Mr. Potato Head, remain among the most sought of all PEZ dispensers. Mint on Card (MOC) Make-A-Face dispensers from Germany held their $2,500 value, slightly less than the American made version, which still commanded an astonishing $3,000 MOC.

Space/Science Fiction

Duplicating their undisputable reign over last year's crop of science-

fiction toys were all things *Star Wars*. Outstanding though it may be, however, Kenner's new lineup of action figures and vehicles showed some signs of overproduction as retail store shelves seemed to display more than enough to satisfy the most insatiable amount of market demand. Fortunately, demand for elusive figures, including Darth Vader with removable helmet, still hovered at high altitudes last year.

Bolstered by exciting new releases like their alluring 9-inch Seven of Nine action figure, Playmates' *Star Trek* lines continued their mission on retail shelves and through the secondary market. A slew of new *Lost in Space* toys found their way into collections as the feature film blasted into theaters, rekindling interest in items from the vintage TV series as well.

Classic Tin

According to Bill Shepardson of Vintage Toys in Missouri, one of the most significant trends among collectors of classic tin and cast-iron toys may very well be due to the generation gap. More and more, Shepardson is seeing collections being sold off as this generally older segment of toy collectors strives to get their estates in order. Auctioned pieces continued to garner astonishing prices as well.

Vehicles

Miniature vehicles of

all kinds kept pace in the toy collecting world last year, as Krause Publications'

What's hot in die-cast? Vintage die-cast like classic Corgi, above, or custom Hot Wheels, left, are highly prized.

new *Toy Cars & Vehicles* magazine could attest.

Classic names like Corgi, Dinky, Hot Wheels and Matchbox held fast to their lanes around the vehicles track in keeping with time-honored tradition.

Perhaps truest to their name, Hot Wheels Treasure Hunt and First Edition cars seemed to generate the most heat among newer issues.

"Hot Wheels and HO slot cars are really booming," said Perry Mohney of the Toy Exchange. "With Hot Wheels it's the newer stuff, you know, people keeping up with the First Edition series and stuff like that. Everybody's trying to stay on top of it, just to keep up with the new stuff coming out; it's nuts."

Mohney stated that some First Editions are "marked up $15 to $20 right out of the chute because you can't find them." Collectors spent last year scooping up the hard-to-get cars — most which sell for under $1 each — in efforts to resell them on the secondary market.

"Even I pay the after-market price on it," admitted Mohney. He added that Hot Wheels Special Editions including their JC

Whitney pieces, the Hot Wheels VW bus and the Service Lube bus (available as special offers from products like Van de Kamps foods) are "good secondary market pieces."

Mattel's 30th anniversary Hot Wheels generated some collector attention, even though the new releases of vintage cars retailed for around $5-$6 each.

"Old Corgi, old Dinky and old Matchbox still sell real well," said Mohney. "Out of all of them, I'd say Matchbox is probably the hottest."

While farm toys have never gone out of style among collectors, an auction at the Old Gold National Antique Tractor Show harvested some serious bids. A boxed six-piece IH-Split Rim 400 Farm Set started at a $4,000 minimum bid and quickly escalated to an astonishing $13,750 as two enthusiasts set off a bidding war between them. The set was sold by Aumann Auctions in Springfield, Ill.

Updating the secondary toy market has become an annual undertaking for contributing writer Guy-Michael Grande, former Toy Shop columnist.

Krause Debuts Online Auctions

if you collect it — you'll find it

COLLECTit™
www.collectit.net .net

Love toys? Like using your computer?

Krause Publications offers several ways for you to find the toys you want for your collection.

An online live auction site is now available through Krause's web site at **www.collectit.net.** The site will function as many other sites do, with prospective buyers logging on, perusing auction listings and making bids.

The auctions will be live, interactive and ongoing. The site will be updated in real time so bidders can follow each auction, and participants can search the histories of sellers before bidding. If they buy an item, bidders will contact a seller directly to finalize details.

Several features incorporated into Krause's auction site will make it unique among online auction services. Among these features are a bidder's passport allowing single sign-on and quick navigation through the site, a multi-bid feature allowing bids to be placed on multiple items from a single screen, notification via e-mail if a bidder has been outbid or won an auction, and user-friendly uploading of graphics and descriptions for sellers offering one or multiple items.

"Selling collectibles online is a perfect match for the Internet. It's quick, easy and doesn't waste time," said Krause Executive Vice President Roger Case.

"The web auction is a phenomenon that is raising online retail to new heights.

"Online auctions infuse the buying process with fun, competition and community, the same as physical auctions. Krause plans to use its extensive knowledge of collecting to create and provide the best site available for collectors," Case added.

"This is the beginning, not the end, of where Krause Publications is going on the Internet," Case said.

"We are looking at continuing to expand our services to hobbyists in a variety of ways, and it will be exciting for enthusiasts to watch our services grow. The larger collectit.net becomes, the more it will establish itself as the place on the Internet to buy and sell collectibles and hobby-related items."

Also at collectit.net is a large classified advertising site linked to most of Krause's 34 periodicals, including *Toy Shop*.

Like what you see? Want to find out more? You can also learn more about Krause's collectibles books and magazines at **www.krause.com**.

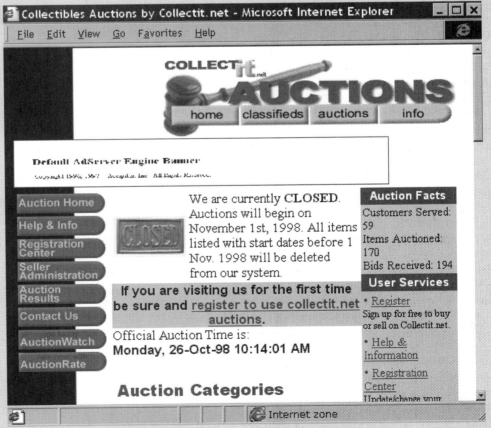

Toy Shop 1999 Annual

39

Welcome to the Toy Shop!

Behind the Scenes at the Hobby's Top Magazine

By Sharon Korbeck

"When I am grown to man's estate
I shall be very proud and great,
And tell the other girls and boys
Not to meddle with my toys."

Those lines from Scottish author Robert Louis Stevenson, although written in the late 19th century, are applicable even today — especially to toy collectors.

Stevenson may have been speaking figuratively, but fans of collectible toys are just as adamant about their treasures.

Krause Publications — a company with an inherently keen insight to the hobby industry — began to recognize in the 1980s a growing interest in collecting old toys.

Krause was already actively involved in the hobby of collecting old cars, and more and more vehicle toys began appearing at car shows.

"The company recognized the emergence of a growing active toy collecting marketplace," said Krause Executive Vice President Roger Case.

Krause's first toy publication, Toy Shop, debuted in August, 1988. The tabloid-size monthly publication was what's known in the publishing industry as a "shopper" or "trader" — an all advertising marketplace to buy, sell and trade. It followed the same approach, Case said, that Krause had taken before in other hobbies. "It was definitely one of our more successful

[magazine] launches," Case said.

Toy Shop was initially designed to target the collectors of toy vehicles. But the baby boomer audience — interested in recapturing the toys of the 1950s and 1960s — became a stronger draw. Adults who grew up in those decades now collect space toys, movie monsters, Barbie, G.I. Joe and TV characters — the toys they remember from their youth.

The first issue of *Toy Shop* had 48 pages and 115 display ads. The magazine's progress was slow, but sure, and it was evident the hobby was picking up speed.

By 1994, the magazine had grown to over 250 pages every month, and the publication changed its frequency to bi-weekly (every two weeks).

The *Toy Shop* editorial staff also produces the paperback *Toy Shop Annual* at the end of each year as a resource directory and recap of the past year in collectibles.

Toys Hit the Newsstand

Toy Shop was a success with hardcore and veteran toy collectors. But

what about those with a casual interest in the nostalgia and growing value of old toys?

To reach more casual collectors, Krause introduced *Toys & Prices*, its first newsstand magazine for toy collectors, in fall, 1992. With a color cover and hundreds of pages of price guide, more than 65,000 were placed on national newsstands.

The following year, Krause acquired the Florida magazine *Toy Collector*. The two magazines were combined in June, 1993, and *Toy Collector and Price Guide* was born. By that time, circulation had hit nearly 70,000 copies, making it the most widely-circulated magazine in the toy hobby.

Toy Collector and Price Guide's time on the newsstand was short, however. A magazine's survival on the national newsstand is often tenuous due to readership, competition and advertising support. *Toy Collector and Price Guide* folded in early 1996.

What's My Toy Worth?

With both *Toy Shop* and *Toy Collector and Price Guide*, Krause recognized it had to address the main question its readers were asking — what's it worth? That's why price guides have truly been the backbone of the Krause Publications lineup. In the toy division, price guides are top priority.

"*Toys & Prices* [the magazine] was always envisioned with the book in mind," said Case, first editor of the magazine. That book, which debuted in late 1993, was also titled Toys & Prices. Since then, Toys & Prices — Krause's annual price guide — has seen six editions. It now nears 1,000 pages of prices for thousands of toys from action figures to Barbie to vehicles and is the top selling general toy price guide in the hobby. *Toy Shop* editor Sharon Korbeck has overseen the past four editions.

KP Acquires the Competition

While *Toys & Prices* was an unqualified success, competition remained. The strongest price guide contender was Richard O'Brien's *Collecting Toys* book, published by Books Americana. When Krause purchased Books Americana in 1996, the O'Brien title was included, enhancing Krause's stall of toy publications. Soon after, O'Brien retired, but his popular book, currently in its 8th edition, is now edited by Krause's toy books editor Elizabeth Stephan.

Additional acquisitions in the book division — especially of Chilton Book Company's Warman's and Wallace-Homestead titles — added literally dozens of toy collecting books to Krause's stable. Titles covered popular hobby areas like board games, Fisher-Price toys, Little Golden Books and character toys.

In addition, Krause has continued to publish single-topic toy collecting books. Among those is the much-respected *The Ultimate Barbie Doll Book* by Marcie Melillo and *The Complete Encyclopedia to G.I. Joe* by Vincent Santlemo.

In With the New

While Krause's success with toy books grew, the company remained active in the magazine area as well.

Collectors liked reading about the toys they collected, and they craved information and values about new collectible toys. But since the demise of *Toy Collector and Price Guide*, Krause lacked a forum for that information.

So in early 1996, *Toy Shop* began including feature stories, news articles and price guides in every other issue.

But with a cover that looked like a catalog, some readers were still confused about the mission and focus of the publication.

Beginning with the August 15, 1997 issue, *Toy Shop* adopted a full page, four-color cover, designed to attract more readers on the national newsstand. *Toy Shop* now covers the hottest new collectible toys and

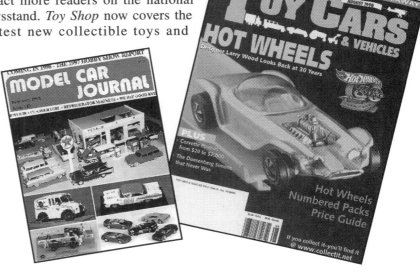

continues to include values for vintage and new toys. Among other vital coverage, editor Sharon Korbeck's annual report on the American International Toy Fair in New York City gives readers a first-hand look at the newest toys of the year.

Toy Cars & Vehicles

One of the largest segments of the toy collecting hobby is vehicle toys — whether its vintage pressed steel Tonka trucks, bold NASCAR die-cast models or classic new Hot Wheels die-cast cars.

In 1998, 10 years after *Toy Shop*'s debut (which was originally targeted at the toy vehicle market), Krause introduced the monthly *Toy Cars & Vehicles*.

Toy Cars & Vehicles grew out of the acquisition of *Model Car Journal,* a small bi-monthly publication aimed at model kit builders. The new magazine expanded its focus to include

columns on die-cast models, farm toys, promotional models and vintage vehicles.

Price guides are also a part of every issue.

In late 1998, after only seven issues, *Toy Cars & Vehicles* grew to a circulation of more than 16,000.

Work Perks . . .
Or What About All Those 'Free' Toys?

By Sharon Korbeck

OK, the answer is "yes, yes and yes."

We do get a lot of toys here in the office.

We do play with them.

We do race our Evel Knievel Stunt Cycle down the hallways and throw Silly Slammers against the wall.

What happens to all those free toys?

We thought you'd like to know, since it seems everyone in our own offices here at Krause Publications is very curious about just that.

C'mon, we're just like you. How would you react to getting boxes of toys in the mail? Let's just say, we love to see what will appear in our mailbox every day. Sometimes its awesome — like Mattel's holiday shipment of the latest and greatest for Christmas.

Sometimes it's humorous, like the week we received a different wacky troll doll almost every day.

And sometimes it's just downright goofy (a battery-powered lollipop that plays music in your mouth??).

Anyway, it's our job to let readers know about these new, innovative and ultimately collectible toys. So to be able to pan or promote them effectively, we need to see, touch and play with them.

OK, there's really no disputing that. But back to the question at hand — what happens to all those toys after we've photographed and written about them?

1. Return them. Some companies require return on items shipped for photography or review. This is especially true with prototype, limited-edition, exclusive or expensive items.

2. Have a contest. Most of the toys offered in giveaways in *Toy Shop* or other magazines come from manufacturers. So the next time you see a great model kit giveaway, you'll know we probably received a lot of model kits.

3. Donate them. Krause Publications participates in a local Toys for Tots holiday drive. Many new toys are donated to that cause.

4. Display them. Our company visitors center features display cases full of collectibles representing our many hobby areas. *Toy Shop* has also participated in toy displays at local libraries.

5. Keep them. Yes, a perk of working where we do is keeping some of the toys. Hey, we're collectors too! Small or inconsequential shipments often remain in our cubicles. Toy Story's Woody peeks out from behind my Rolodex. A six-inch figure of *The X-Files*' David Duchovny is positioned right where I can see him every single morning (what a way to start the day!). A plush Wienermobile sits comfortably next to a small Yoda shrine on my computer. And don't even ask what Spawn is doing to a Teenie Beanie Baby on my bookcase.

So, now you know. You can be jealous, though; we realize we have fun jobs. And while we're willing to take the perks, we're also willing to do the work.

Sure beats my last job. I worked in the funeral service industry . . . and you don't even want to know about those freebies!

Sharon Korbeck is editor of Toy Shop *and* Toy Cars & Vehicles *magazines. Her stash of toys on her computer currently includes a South Park squeezie toy, Mr. Potato Head, a Weeble, Kyle Petty car and Toy Story's Crawling Baby Face.*

How Many People Does it Take To Put Together a Toy Magazine?

About a dozen people, in addition to support staff throughout Krause Publications, keep Krause's toy division on schedule.

Publisher Mark Williams oversees the toy publications. Kevin Novak, toys advertising sales manager, oversees sales representatives Rhonda Hainzlsperger, Chuck Lamers, Dawn King and Norma Jean Fochs. Cindi Phillips and Dawn Loken are the sales department assistants.

The magazines' editorial staff — responsible for writing, editing, layout and design — is comprised of Sharon Korbeck, editor; and Mike Jacquart and Merry Dudley, associate editors. Graphic artist Chris Pritchard designs the covers for *Toy Shop* and *Toy Cars & Vehicles*. Elizabeth Stephan is the book editor for the toys division.

Toys That are Just Plain Silly!

Gibson Greetings' Silly Slammers are Office Hits

By Mike Jacquart

It's important to leave valuable toys unplayed with. But it's equally important to *play* with other toys.

One such item is Gibson Greetings' Silly Slammers. Silly Slammers, called America's first "beanbags with an attitude," are talking characters with outrageous expressions, bright colors and funny phrases. Each miniature beanbag blurts out a unique phrase or phrases when dropped or slammed.

Over 10 Million Silly Slammers

Since the introduction of the original line of eight Silly Slammers in September 1997, Gibson has sold well over 10 million Silly Slammers, according to Gary Rhodes, Gibson's director of corporate communications.

The demand has also caught the eye of national broadcast and print media, with Silly Slammers featured on shows airing on ABC, Disney and Fox and in stories in *USA Today* and *The Wall Street Journal*.

The Original Eight Slammers

Original Silly Slammers include B. Earp, who belches; the lushly-lipped Polly Pucker Up, who delivers a long, loud smooch; the English rocker Crash, who delivers a scorching electric guitar riff and Botch, who

ABOVE LEFT: "That was a practice swing!" "Have you considered bowling?" are phrases any struggling golfer can relate to. ABOVE RIGHT: You probably know someone in an office that can relate to Buddy Brown Noser (glasses) or Mr. Excuses (smile). Gibson Greetings' Silly Slammers like these are inexpensive (approximately $7 each), fun and are great stress relievers!

RIGHT: Silly Slammers IV includes the bearded Frisco ("No problem," "Bummer!" "Duuude!"), the ponytailed Harriet ("Get real," "Get a life!" "Over it!") and the frown-faced Fava ("Why I oughtta . . .!" "And for the love of . . .!" "What the . . .!)

exclaims "oh no!" According to Gibson chairman, president and CEO Frank O'Connell, Silly Slammers ". . . make great collectibles for kids, stress relievers for adults and are a fun way for just about anyone to express themselves. . ."

How it All Began

What are the origins of these unique items that seem to bring a smile or laugh to all who drop or slam them?

Bob Cavellier, gifts/plush strategic business unit leader at Gibson Greetings, manufacturers of the popular plush beanbags, said Silly Slammers resulted from Gibson's desire to make other products instead of just greeting cards and related items.

While not sure who coined the actual name of the toys, Cavellier credits his boss, Greg Ionna, executive vice president of marketing, with coming up with the idea for Silly Slammers. Cavellier recalled Ionna saying that Gibson needed to come up with "the next Beanie Baby." Ionna drew a face on a simple beanbag, and hair and legs were added, "but we didn't know what we had at the time," Cavellier said.

The Unveiling at Toy Fair

Gibson officials unveiled 70 Silly Slammers designs at the American International Toy Fair in New York City in February 1998. It was an idea

The original eight Silly Slammers introduced in fall 1997.

that definitely caught the attention of people in the toy industry, Cavellier said.

"It was like, 'This is Gibson?'" Cavellier recalled. Still, Cavellier said it wasn't easy for the 150-year-old greeting card company to gain admittance to the premiere event that brings together the biggest manufacturers and buyers in the *toy* industry.

Cavellier credits Doug Guendel with helping Gibson land appointments at Toy Fair. Guendel, vice president and general manager of Gibson's entertainment sales division, brought sales representatives with him to Gibson from Golden Books, where Guendel previously served as a vice president.

Promotions Stepped Up

Merchandising and marketing plans have since stepped up awareness of the miniature beanbags. Cavellier noted his company started a national advertising campaign promoting Silly Slammers that was slated to run through Christmas.

The Silly Slammers III line consists of, clockwise from top left: Yada ("Yada yada yada" "Whatever!"), Jammer ("Later!" "Adios" "Bu bye now!"), Penelope ("As if" "Yeah — right!" "No way"), Tiffany ("Loser!" "Whatever" "I don't think so") and Lloyd ("Hello" "You're bothering me!"). Lumpy, the Silly Slammer pictured in the middle, says "Help me help me!" and "Oh no!"

Silly Slammers are a big hit at Krause Publications. Lisa Wilson, pictured with her lookalike, is one enthusiast.

More Sounds — Even Lights

In addition, while the first wave of Silly Slammers could only recite one phrase, Slammers issued since then can exclaim four different phrases. Lights and sound effects are also being added to the beanbags. They're available at stores like Toys R Us, 7-Eleven, Winn Dixie, Kroger and Harris Teeter.

Diverse Lines

Other lines include an Office Group, Summer Sports, Slammers III, Slammers IV, Fitz & Tantrums, Bugs, Wrestlers, Collegiate Sports and Christmas.

The Office Group features five characters that strike a familiar chord with office workers all over the country — including Krause Publications, where the beanbag toys are enjoyed by many employees.

Personalities include The Boss ("You're fired!" "Blah blah blah!" "I need it now!"), the nervously-sweating Mr. Excuses ("I can explain!" "I wasn't even there" "My secretary lost it"), Ms. Administrative Assistant ("You're welcome" "That's a good one!" "You want it when?") and bespectacled

Toy Shop *Editor Sharon Korbeck surrounded by a plethora of zany Silly Slammers.*

Fitz Gerald (middle), part of the Fitz & Tantrums Silly Slammers series, says "Yahhhh!" and "Owwww!" Other Slammers in this line, pictured clockwise from above left are: Fitz Simmons ("I'm going crazy" "Augh!"), Chaz the Spaz ("Nyah nyah nyah" "Cuckoo cuckoo"), Ima Whiner ("Boo hoo hoo" "No fair!"), Julius Seizure ("I won't I won't" "Bl-l-l-ib!") and Fitz Patrick ("No no no!" "Braaa!")

Buddy Brown Noser ("How *do* you do it?" "Oh, I agree!" "Pure genius!") and a computer character that makes a crashing sound and says "Goodbye files."

Cavellier said the office line was developed "after receiving an enormous response from adult consumers who said they were buying our earlier generations of Silly Slammers for their offices and work areas."

Halloween Silly Slammers include Mummy ("I want my mummy" "I'm petrified!"), Werewolf ("Happy Howlloween!" "Howl!!!!") and a Green Monster that resembles Frankenstein ("Boo hoo, I'm so scary!")

Christmas beanbags include an Elf ("I Quit!" "Christmas again? Try gift certificates"), a Snowman ("Turn down the heat" "I'm melting!") and a present ("Gee, thanks" followed by a crashing sound "Oh, you shouldn't have.")

Who comes up with the ideas? How are they made?

While consumers generally pay about $7 for one Silly Slammer, bear in mind it costs several thousand dollars to produce a single design from start to finish. Toss in the millions it takes to promote a product via advertising campaigns and through the media, and it's obvious a *lot* of Silly Slammers have to be sold before Gibson can turn a profit.

Not that Cavellier is complaining. High overhead is the nature of any toy line — besides, how many toys are this much fun?

So fun that Cavellier said it's not unusual for him to wake up at 3 a.m. to jot down an idea for a new Silly Slammer.

"We watch the market," he said. "What movies will create trends? What are toy companies doing? We try to get trendy sayings out while they're still trendy."

For instance, one character, named Yada, features the immensely-popular catch phrase "yada, yada, yada."

How Do They Talk, Anyway?

Surprisingly, only a handful of workers at the Cincinnati-based Gibson Greetings work on Silly Slammers.

"We come up with the ideas, but outsiders [other firms hired by Gibson] tighten up the drawings," Cavellier explained. Paradigm Studio of Cincinnati provides and records the voices, he added.

Silly Slammer "voices" are placed on an impact activated sound module — a fancy name for a simple device sewn into the Slammers. The inside of a Slammer also contains a round battery. In a nutshell, these two

simple devices are what makes a Slammer "talk."

Features are embroidered onto a piece of plush material. Actual manufacturing takes place in Asia.

What's in Store Next?

Future Silly Slammers will include a Smart Guys line that will feature memorable artists and scientists like Mozart, Shakespeare, Einstein, Freud and a composer Bach that will say "Oh, my aching Bach!"

Silly Slammers enthusiasts should also watch for a college series. For instance, a University of Michigan Slammer sings "Go Blue!" and the school's fight song.

On the heels of the animated *Antz* and *A Bug's Life* movies, a Bugs Slammers line is also underway.

In addition, talking key chains can make a trip in the car more enjoyable. T-shirts and other items also will join the expanding Silly Slammer universe, Cavellier said.

Licensed Silly Slammers like an *Austin Powers* series is also in the works — but don't expect to see a do-it-yourself Slammer (one in which you could record a saying yourself)

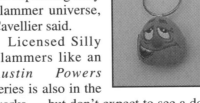

ABOVE: An impact activated sound module — a fancy name for the device (plus a battery) that makes a Silly Slammer talk.
ABOVE RIGHT: Key chains are now part of the Silly Slammers line-up, and other items are planned.

1999 Toy Shop Annual *and* **Toy Shop** *cover designer Chris Pritchard (above and left) juggles some Silly Slammers.*

ABOVE: The Silly Slammers Christmas line includes an elf, Santa Claus and a snowman. BELOW: More of a Halloween enthusiast? Try these Slammers.

anytime soon.

"We're entertaining that idea," Cavellier said, "but that's something we'll have to think about."

Regardless,

it's clear the sky's the limit as far as these incredibly versatile toys are concerned. As Cavellier states, while ideas for Silly Slammer lines abound, "it's a good problem to have!"

Happy Anniversary!
Hot Wheels, Harley Both Celebrated Birthdays in 1998

By Christina Bando

What's the one thing every person, place or thing has in common? The one thing that occurs once every year, for everyone — birthdays and anniversaries, of course.

The toy industry is no exception. Whether they are called birthdays or anniversaries, they are a cause for celebration, and many companies celebrated in a big way in 1998.

From Hot Wheels to Harley-Davidson to NASCAR, some of the biggest and best toys turned another year older in 1998 and caused celebration across the country. So grab some cake, ice cream and a hat and join the party!

Mattel Hot Wheels cars turned 30 in 1998.

100 YEARS

Gund

Nearly everyone, at one time or another, has owned a stuffed teddy bear. Teddy bears seem to have been around forever.

When the teddy bear came into existence, Gund was there to design and produce it in the early 1900s. And after an amazing 100 years in the toy industry, Gund is still going strong with the release of its 34-inch soft mohair centennial bear. This is no ordinary bear; keeping with the 100-year theme, the handmade bear, designed and signed by director of design Rita Swedline Raiffe, was auctioned for $100,000!

Gund was founded in 1898 by a German immigrant, Adolph Gund, who took his toy making very seriously. When a toy buyer came to Gund for a teddy bear, he spent nearly an entire night with a few yards of

plush, designing four different sizes of teddy bears.

Because of Gund's early efforts, the company began to flourish by 1914. The company also became a pioneer in the toy industry, being one of the first to produce animal versions of cartoon characters. These characters included Felix the Cat in 1927, as well as Mickey Mouse, Minnie Mouse, Pluto and Donald Duck in the 1940s.

95 YEARS

Crayola Crayons

From toddlers to preschoolers to college kids, Crayola crayons have made their mark on the coloring books, refrigerators, furniture and lives of many. It has been 95 years since the birth of an "oily stick of color," and since then, nearly 100 billion Crayola crayons have been made and worn down by children everywhere.

The trade name Crayola was coined in 1903 by Alice Stead Binney, the wife of crayon producer Edwin Binney. The name's roots lie with the French word *craie*, meaning stick of color; and the word "oleaginous," meaning oily. Since its conception, the word "Crayola" has become synonymous with crayon, making the two together a household name.

Binney & Smith, maker of Crayola products, continues to produce over two billion crayons every year, much to the delight of children and crayon enthusiasts.

Harley-Davidson Motorcycles

Dressed in black leather and straddling a "hog," motorcycle riders and collectors have helped make the terms "Harley-Davidson" and "motorcycle"

Would you pay $100,000 for a teddy bear? A Japanese collector recently did. Proceeds from Gund's one-of-a-kind $100,000 bear — that noted its 100th anniversary — went to four charities that benefit children.

nearly inseparable. The year 1998 marked the 95th anniversary of this motorized two-wheeler that has swept the nation since its birth.

Along with the Harley-Davidson motorcycle, the toy replicas began to pop up in collections almost as soon as the hogs hit the road. In the toy industry, Hubley began to offer an extensive line of cast-iron Harley-Davidson replicas throughout the 1930s.

When the motorcycle received attention from the military, toy manufacturers were right there, redesigning their product to fit the trend. Cast-iron toy motorcycles began to feature military and police riders after 1916, when Harley-Davidson motorcycles were used to track Pancho Villa and his band in Mexico.

Hubley was not the only toy company to manufacture motorcycle toys, however. The Louis Marx company produced some of the nicest tin American motorcycle toys. And not all motorcycle toys were made out of metal; in the 1950s and 1960s, plastic and rubber motorcycle toys became the rave. By the time the 1990s rolled around, Matchbox had entered the Harley-Davidson cult world with its extensive line of licensed Harley-Davidson die cast.

85 YEARS

Erector Sets

A relatively old-timer in the toy industry is re-emerging on its 85th anniversary. The Erector set was one of the most popular toys in America in the 1950s and 1960s. After a 20-year downturn in sales and more than 10 years off the American market, the Erector set is once again building up attention with its high quality and booming sales.

The Erector set began its legendary status with the A.C. Gilbert Co., a pioneer in the area of scientific and educational toys. Long before they were fashionable, educational sets found a hero in the A.C. Gilbert Co. With sets in chemistry, optics, magnetism, astronomy and other sciences, these toys challenged the minds of many young children who had high hopes of becoming scientists one day.

The Erector set was no exception.

Nearly every household in middle-class America in the last several decades had an Erector set. What started out in 1911 as a passing thought during a train ride turned into one of the most popular toys in America.

Designed by A.C. Gilbert, the set began as a prototype of steel girders and evolved into a nationwide craze. When the 1913 New York Toy Fair rolled around, Gilbert was ready to present the first Erector set to the world. In response, he was given more orders than his small company could fill. By 1915, the Erector set had won the Gold Medal at the Panama-Pacific Exposition and had sales of more than 30 million sets.

After A.C. Gilbert's retirement in 1956, Erector sets lost popularity and were eventually pulled from the American market by 1988. They have returned recently, however, and collectors and children alike have embraced them with open arms.

65 YEARS

Popeye

Long before the Mel Gibsons of today and the Marilyn Monroes of yesterday, Popeye the Sailor man was making his debut on the silver screen. Popeye is still entertaining audiences on his 65th birthday as a film star, with the release of a home videotape compiled from eight black-and-white, 1930s-'40s cartoons.

Popeye's original role at his conception was one of a throwaway character in Elzie Crisler Segar's Thimble Theatre. However, the minute Popeye strutted his stuff on screen with co-star Betty

Everyone's favorite sailor turned 65 in 1998. How many cans of spinach do you think he's eaten over the years?

Superman is in great shape for age 60, and many older collectibles touting his likeness are extremely valuable.

Boop, American audiences young and old fell in love.

The toy manufacturers were quick to follow suit, releasing literally truckloads of items to the Popeye-frenzied audience. Soon, Popeye found his way to the printed page, popping up in comic books and newspapers alike.

By 1935, the Popeye comic strip was printed in over 300 U.S. newspapers and carried in 12 foreign countries. Today, printed Popeye material can be found all over, in practically every language in the world. Popeye has traveled a long way from his humble Chester, Illinois, beginnings.

With reference to toys, the most treasured items sought today are the "tins" from the 1930s. These wind-ups feature Popeye doing everything from boxing to dancing. Today, prices have skyrocketed so much for these coveted items that even the boxes they were sold in are worth hundreds of dollars apiece.

60 YEARS

Superman

After six decades of beating the bad guys, the man of steel has definitely proven to be indestructible. Even after the death of his creator, Jerry Siegel, in 1996, and his own death at the hands of Doomsday in the early 1990s, Superman has bounced back into immortality in comic books, cartoons and action figures, among others.

Now, at age 60, Superman is still doing what he does best — saving the world from evil . . . and selling thousands of collectibles. The incredible success of Superman was a result of good timing by his creator. The man of steel emerged in the late 1930s, when America was going through the Depression and desperately needed a hero.

His motto of "truth, justice and the American way" helped contrast the fascism in the age of World War II's Adolf Hitler. Those ideals have stuck with Superman throughout the years, and his American hero status has remained intact.

Superman had his debut in June, 1938 in *Action Comics #1*, but he appeared five years earlier in a short story Siegel had written for the magazine *Science Fiction*. The story, titled "The Reign of Superman," depicted Superman as a man who gained and misused fantastic mental powers. A far cry from the indestructible American hero of today.

Although most widely known through comic books, Superman has emerged in other toys as well. Early Superman comic books and premium rings are among the most valuable collectibles today.

50 YEARS

Scrabble

How many readers remember sitting around the family dinner table on a quiet evening at home, arranging little wooden tiles on a cardboard game board, trying to come up with the most outrageous word worth the most points?

With a 50-year resume of educational fun and family entertainment, Scrabble is still one of the most popular games ever invented. With a following of over 33 million enthusiasts, it is no wonder Scrabble can be found in one of every three American homes.

The Great Depression indirectly gave birth to the idea for Scrabble. Alfred Butts, the father of Scrabble, was an unemployed architect from Poughkeepsie, New York. Because of his interest in anagrams and his available free time, Butts was able to tinker

around with the game for more than a decade, finally developing the officially trademarked Scrabble game in 1948.

Today, more than 100 million Scrabble games — now made by Hasbro — have been sold worldwide since its conception.

NASCAR

Could you imagine a 50-year-long race to the finish line? NASCAR — The National Association of Stock Car Auto Racing — has been speeding by competition for half a century now and has yet to reach the end of the race.

What began as a tiny planning meeting at a Daytona hotel in 1947 has exploded into one of the most exciting stock car associations ever conceived.

In honor of the 50th anniversary, many toy companies have released die-cast versions of classic NASCAR vehicles.

Marvin the Martian

After 50 years of being very angry at earthlings, Marvin the Martian hasn't changed a bit. This Warner Brothers' space creature still totes around his disintegrator, and he is still trying to destroy the earth.

45 YEARS

Fort Apache

Who doesn't remember playing Cowboys and Indians as a child?

This game has been passed down from generations, spanning all the way back to the actual frontier fighting. Louis Marx, after recognizing the popularity of the game, created what was soon to become the single most popular play set of all time — Fort Apache.

After 45 years, this play set is still a prized possession of any collector or child. The set was so popular, its production run spanned several decades, and many variations were

produced over a 20 year span. One such variation was an official licensed Rin Tin Tin Fort Apache.

40 YEARS

Paddington Bear

Michael Bond, legendary British children's author, created a legacy that has lived four decades after its conception. Paddington Bear, who turned 40 in 1998, is a proud part of that legacy. To aid in the celebration, Eden, the licensee of Paddington Bear, released a special 40th anniversary Paddington bear.

35 YEARS

Charmin' Chatty

Back in 1963, this lanky little girl with her granny glasses chatted her way onto the cover of *Saturday Evening Post*, a feat that even Barbie would envy. Designed as a pullstring doll with a twist, Charmin' Chatty came with five insertable records that gave her over 120 varied responses.

Other play sets were also available, giving Chatty the "right words to say" for any occasion, from birthday parties to shopping to playing nurse. A Travels 'Round the World play set even allowed Chatty to speak in seven different languages with a simple pull of the string.

Recognize Marx's Big Loo from 1963? The 35-year-old robot is worth over $2,000 in Mint in Box condition!

Big Loo

The space-age robotic "friend from the moon," Big Loo, was perhaps one of the most versatile toys made in the 1960s. Children could spend hours shooting darts out of his chest, balls from his arms, and multi-finned rockets from his foot. And if boredom set in with his aggressive features, children turned to Big Loo's built-in scanner scope, whistle, bell, compass, and Morse code clicker.

Produced by Marx in 1963, Big Loo retailed at a measly $10 and was the biggest and best spaceman money could by. After 35 years, Big Loo is still his homely, blinking-red-eyes self, but he carries a much heftier price tag. A Mint in Box Big Loo would cost a collector $2,200.

Midge

It seems a year can't go by without at least one Barbie anniversary. Mattel's Midge doll, advertised as a friend of Barbie, was looking as stylish as ever when she turned 35. Mattel released a 30th anniversary edition of the doll in 1998.

Mouse Trap

After celebrating 35 years of rodent-catching, Ideal's Mouse Trap Game is not the least bit "rusty." Designed in 1963 by Marvin Glass, this colorful, three-dimensional strategy game has trapped children's fancies for over three decades. The game was so popular when it first came out, Ideal sold 1.2 million copies within the first year!

Easy-Bake Oven

Much to the delight of children, the Easy-Bake Oven first appeared on the market in 1963 for the retail price of $15.99. After some pretty sweet sales of more than 500,000 units that first year, the oven became the hottest selling girl's toy since the creation of dolls.

Since that original run 35 years ago, the Easy-Bake Oven has expanded to include a wide variety of tasty treat-creators, including a potato chip maker, a blender/juicer, a taffy machine, an ice-cream and frozen treats maker, among others. Thanks to the wide variety of appliances and the consumers' love of sweets, more than 16 million Easy-Bake Ovens have been sold and more than 100 million mix sets have been eaten! How's that for a sweet tooth?

30 YEARS

Hot Wheels

Parents have been giving their children toy cars for decades, whether it be for Christmas, birthdays or any other conceivable holiday. Strange thing is, a majority of children received their first car before the age

Holy lasagna! That fat cat Garfield is 20! Garfield is pictured with his creator, Jim Davis.

of seven. In fact, by the time these youngsters finally turned 16, they had already accumulated sometimes over 50 cars. How is this possible? Two words . . . Hot Wheels.

Thanks to the husband and wife team of Ruth and Elliot Handler, along with Harold Matson, one of the best known toy companies around, Mattel, was formed in 1945.

A little more than 20 years later, through the creative genius of Jack Ryan, Harry Bradley and Ira Gilford, the first line of die-cast Hot Wheels was born. And man, were they hot.

Within the first year of business, Hot Wheels experienced sales that were seven times what was expected. Soon, these tiny cars raced across the country into the living rooms and hearts of America.

On the 30th anniversary of the first Hot Wheels line, Mattel offered a 30th anniversary line including the early Twin Mill. The special edition set contained a car from every year since 1968, from the '68 Deora all the way to the '97 Scorchin' Scooter.

20 YEARS

Garfield

On June 19, 1978, a champion to lasagna platters everywhere was born. Garfield, the fat, lazy, coffee-drinking, dog-punting cat, managed to sink his claws into 41 different U.S. newspapers the first day of his conception.

And after 20 years of hating Mondays, drinking coffee and adoring lasagna, Garfield is still entertaining with his dog-punting, spider-whacking antics. Not only is he the star of his own cartoon, but his face is plastered over the comic strip pages of over 2,500 newspapers.

Jim Davis, the father of the famous cat, began his legacy in the basement of his home in Muncie, Indiana. With the help of his wife, Davis eventually established PAWS in 1981 to keep up with the demand for Garfield products. In fact, Garfield had grown so popular that by the time 1989 hit, a 36,000 square foot building was needed to house the fat cat, along with 55 employees.

So what's in store for the master of meow? A Garfield theme park, of course. The theme park, located in Hendricks County, Indiana, will open its doors in 1999 and will feature rides, attractions, an amphitheater, shops and dining, all in honor of America's best known cat.

15 YEARS

Cabbage Patch Kids

Mattel had a busy year trying to keep up with all the anniversary and birthday celebrations. In addition to the 30th Anniversary of Hot Wheels, Mattel also found its hands full with the 15th anniversary of Cabbage Patch Kids.

Remember 1983? It was the year Cabbage Patch Kids began sprouting in homes across the country. To celebrate its 15th anniversary, Mattel released a large assortment of anniversary dolls including a replica of the one that started it all.

10 YEARS

Toy Shop

Ten years ago, when the toy collecting hobby was just heating up, Krause Publications introduced the monthly *Toy Shop* magazine, its indexed toy publication. Starting as a "shopper" with hundreds of advertisements for collectible toys, *Toy Shop* has grown into a bi-weekly publication including feature stories, news items and price guides.

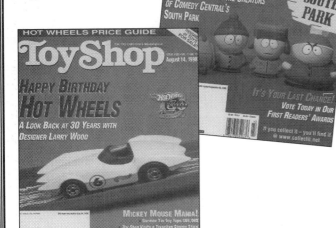

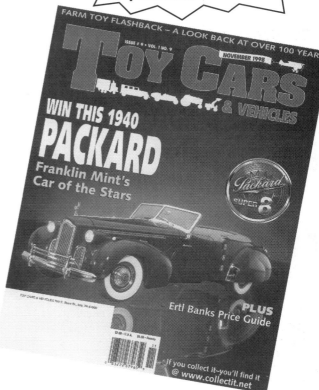

Manufacturers Directory

Have you ever wanted to reach a toy manufacturer but didn't know how? This list can help you.

While not all-inclusive, this list includes many notable toy and toy-related companies currently in business. Remember, many companies will not comment on the secondary market values of their toys. Comments and questions should generally be directed to the customer service department.

21st Century Toys
Action figures
2037 Clement Ave.
Alameda, CA 94501-1317
510-814-0719
www.21stcenturytoys.com

Action Performance
Die-cast cars
4707 E. Baseline Rd.
Phoenix, AZ 85040
602-337-3824
www.action-performance.com

Aladdin Industries
Lunch kits
703 Murfreesboro Rd.
Nashville, TN 37210
615-748-3292

Alexander Doll Company
Dolls
615 W. 131st St.
New York, NY 10027-7982
212-283-5900
www.alexanderdoll.com

ALPI International
Miscellaneous
1186 63rd St.
Oakland, CA 94608
510-655-6456

Applause
Plush, PVC figures
merged with Dakin in 1995
6101 Variel Ave.

Woodland Hills, CA 91365-4183
818-992-6000

Bachmann Industries
Toy trains, planes
1400 E. Erie Ave.
Philadelphia, PA 19124
215-533-1600

Bandai America
Action figures
12851 E. 166th St.
Cerritos, CA 90701
310-926-0947
www.bandai.com

Basic Fun
Key chains
1080 Industrial Hwy.
Southampton, PA 18966
800-662-3380
www.basicfun.com

Binney & Smith
Crayola products, Silly Putty
P.O. Box 431, 1100 Church Lane
Easton, PA 19042
610-253-6271
www.crayola.com

Cadaco
Games
founded 1935, previously known as Cadaco-Ellis
4300 W. 47th St.
Chicago, IL 60632-4477
312-927-1500

Craft House Corporation
Lindberg model kits
328 N. Westwood Ave.
Toledo, OH 43607-3343
419-536-8351

Danbury Mint
Die-cast vehicles
47 Richards Ave.
Norwalk, CT 06857-0001
800-282-9412

Darda, Inc.
Vehicle toys, etc.
1600 Union Ave.
Baltimore, MD 21211-1917
410-889-1023

Duncan Toys
Yo-yos
15981 Valplast Rd.
Middlefield, OH 44062
216-632-1631
www.yo-yo.com

Eastwood Automobilia
Die-cast vehicles, banks
distributor of die-cast vehicles, banks
Box 3014, Dept. PR
Malvern, PA 19355-0714
800-343-9353

Effanbee Doll
Dolls
19 Lexington Ave.
East Brunswick, NJ 08816
732-613-3852

Empire Industries
Miscellaneous
acquired Buddy L in 1995
5150 Linton Blvd.
DelRay, FL 33434
407-498-4000

Ertl Company
Die-cast vehicles, farm toys, banks
1945-present
Highways 136 & 20, P.O. Box 500
Dyersville, IA 52040-0500
319-875-5607
www.ertltoys.com

Exclusive Toy Products
Action figures
2029 Century Park East, Suite 2880
Los Angeles, CA 90067
310-284-5477
www.exclusivepremiere.com

First Gear
Die-cast models
P.O. Box 52
Peosta, IA 52068-0052
319-582-2071

Fisher-Price
Plastic preschool toys, vintage wood pull toys
1930-present; division of Mattel
636 Girard Ave.
East Aurora, NY 14052
716-687-3449

Franklin Mint
Die-cast vehicles
Franklin Center, PA 19091-0001
800-523-7622

Full Moon Toys
Action figures
1645 N. Vine St., 9th Floor
Los Angeles, CA 90028
877-315-6666
www.fullmoontoys.com

G Whiz Enterprises
Lunch box repros
formerly Wonder Planet
14732 Lull St.
Van Nuys, CA 91405
626-683-9200

Galoob Toys
Action figures, Micro-Machines
bought by Hasbro in 1998
500 Forbes Blvd. S.
San Francisco, CA 94080
415-873-0680
www.galoob.com

Gibson Greetings
Silly Slammers
2100 Section Road
Cincinnati, OH 45237
513-841-6600

Graphitti Designs
Action figures
1140 N. Kraemer Blvd., Unit B
Anaheim, CA 92806-1919
800-699-0115

Gund
Plush

1898-present
1 Runyons Lane, P.O. Box H
Edison, NJ 08818
908-248-1500
www.gund.com

Hallmark
Kiddie Car Classics die-cast models
2525 Gillham Rd.
Kansas City, MO 64108-2622
816-274-8519

Hartland Plastics/Steven Manufacturing
Western/sports figures
224 E. Fourth St.
Hermann, MO 65041
314-486-5494

Hasbro Toy Group
Miscellaneous
1940s present; previously known as Hassenfeld Bros.; includes Kenner, Milton Bradley, Parker Brothers, Tonka, Playskool, Galoob
1027 Newport Ave.
Pawtucket, RI 02862-1059
401-727-5582
www.hasbro.com

Jaymar
Miscellaneous
1930s-present, former Marx subsidiary
47 Scrantom St.
Rochester, NY 14605-1054
716-454-7050

Just Toys
Miscellaneous
50 W. 23rd St., 7th Floor
New York, NY 10010-5205
212-645-6335

Kenner Products
Action figures, G.I. Joe
division of Hasbro
615 Elsinore Pl.
Cincinnati, OH 45202
513-579-4927
www.hasbro.com

Krause Publications
Hobby magazines, books (*Toy Shop*, etc.)

1952-present
700 E. State St.
Iola, WI 54990
715-445-2214
www.krause.com
www.collectit.net

Larami Corp.
Miscellaneous
340 N. 12th St.
Philadelphia, PA 19107-1123
215-923-4900

Legends In 3 Dimensions
2032 Armacost Ave.
Los Angeles, CA 90025
310-442-0156

LEGO Systems
Construction toys
555 Taylor Rd., P.O. Box 1600
Enfield, CT 06083-1600
800-243-4870
www.lego.com

Lionel Trains
Electric trains
1900-present
50625 Richard W. Blvd.
Chesterfield, MI 48051
810-949-4100
www.lionel.com

Little Tikes
Preschool toys
division of Rubbermaid
2180 Barlow Rd.
Hudson, OH 44236
216-650-3000
www.rubbermaid.com

Maisto International
Die-cast vehicles
7751 Cherry Ave.
Fontana, CA 92336
909-357-7988

Majorette Toys
Die-cast vehicles
2898NW 79th Ave.
Miami, FL 33122
305-593-6016

Marklin
Toy trains

distributor of German trains
16988 W. Victor Rd., P.O. Box 510559
New Berlin, WI 53151-0319
414-784-1095

Marx Toy Company
Miscellaneous
modern reincarnation of famed Louis Marx toy Company
249 E. Georgia Ave.
Sebring, OH 44672
330-938-8697

Matchbox Collectibles
Die-cast Vehicles
6000 Midlantic Dr.
Mount Laurel, NJ 08054
609-840-1511

Mattel Toys
Barbie, Hot Wheels
1945-present; acquired Tyco in 1997; acquired Pleasant Company in 1998
333 Continental Blvd.
El Segundo, CA 90245-5012
310-252-2000
www.mattel.com

McFarlane Toys
Spawn, other action figures
15155 Fogg St.
Plymouth, MI 48170
313-414-3500
www.spawn.com

Meccano-Erector
Erector sets, construction toys
15 E. 26th St., Suite 1617
New York, NY 10010
212-213-9313

Milton Bradley
Games
1860-present; division of Hasbro
443 Shaker Rd. E
Longmeadow, MA 01028-3149
413-525-6411

Minichamps USA
Die-cast vehicles
bought by Action Performance in 1998
14260 SW 136 St. Bldg #7

Miami, FL 33186
305-253-0452

Moore Action Collectibles
Action figures
3038 SE Loop 820
Fort Worth, TX 76140
817-568-2620
www.mooreaction.com

Nintendo of America
Video games
4820 150th Ave. NE
Redmond, WA 98052-5111
206-882-2040
www.nintendo.com

Nylint
Vehicle toys
1946-present
1800 Sixteenth Ave.
Rockford, IL 61104-5491
815-397-2880

OddzOn Products
Miscellaneous
purchased by Hasbro in 1997
240 E. Hacienda
Campbell, CA 95008
408-866-2966

Ohio Art
Etch-a-Sketch, vintage tin toys
One Toy St.
Bryan, OH 43506
419-636-3141

Parker Brothers
Games
1880s-present; division of Hasbro
50 Dunham Rd.
Beverly, MA 01915
617-927-7600

PEZ Candy
Candy dispensers
founded in 1927 in Austria
35 Prindle Hill Rd.
Orange, CT 06477
203-795-0531

Playing Mantis
Johnny Lightning die-cast, model kits, action figures
3600 McGill St.

Suite 300
P.O. Box 3688
South Bend, IN 46619-3688
219-232-0300
www.playingmantis.com

Playmates Toys
Action figures
611 Anton Blvd. #600
Costa Mesa, CA 92626
714-428-2000
www.playmatestoys.com

Playmobil USA
Figures, play sets
22-E Nichols Ct.
Dayton, NJ 08810
908-274-0101
www.playmobil.com

Playskool
Preschool toys
division of Hasbro
1027 Newport Ave.
Pawtucket, RI 02861-2539
401-726-4100

Poof Toy Products
Foam toys
bought James Industries (Slinky) in 1998
45400 Helm St.
Plymouth, MI 48170
313-454-9552

Pressman Toy
Games
1920s-present
745 Joyce Kilmer Ave.
New Brunswick, NJ 08901
732-545-4000

Racing Champions
Die-cast vehicles
800 Roosevelt Rd. Bldg. C
Glen Ellyn, IL 60137-5839
630-790-3507
www.racingchamps.com

Radio Flyer
Wagons
6515 West Grand Ave.
Chicago, IL 60707
800-621-7613
www.radioflyer.com

Reeves International
Distributor of Corgi, Breyer horses
14 Industrial Rd.
Pequannock, NJ 07440
201-694-5006

Remco Toys/Azrak-Hamway
Miscellaneous
1107 Broadway, Room 808
New York, NY 10010-28022
212-675-3427

Rendition Figures
Action figures
16519 Wildnerness Rd.
Poway, CA 92064
619-592-6866

Revell-Monogram
Model kits
8601 Waukegan Rd.
Morton Grove, IL 60053
708-966-3500
www.revell-monogram.com

Road Champs
Vehicle toys
7 Patton Dr.
West Caldwell, NJ 07006-6404
201-228-6900

Russ Berrie & Co.
Plush
111 Bauer Dr.
Oakland, NJ 07436
201-337-9000

Schuco/Lilliput Motor Co.
Vehicles, tin
P.O. Box 447
Yerington, NV 89447
702-463-5181

Sega of America
Video games
255 Shoreline Dr.
Redwood City, CA 94065
415-508-2800
www.sega.com

Smith-Miller
Vehicle toys
P.O. Box 139
Canoga Park, CA 91305
818-703-8588

SpecCast
Die-cast vehicles
428 6th Ave. NW
Dyersville, IA 52040-1129
319-875-8706

Steiff USA
Plush, teddy bears
founded in 1880 in Germany
200 Fifth Ave., Suite 1205
New York, NY 10010
212-675-2727

Tamiya America
Miscellaneous
2 Orion
Aliso Viejo, CA 92656-4200
714-362-2240
www.tamiya.com

Thermos Co.
Lunch kits
Rt. 75 East
Freeport, IL 61032
815-232-2111

Thinkway Toys
Disney products, miscellaneous
8885 Woodbine Ave.
Markham, Ontario Canada L3R 5G1
905-470-8883
www.thinkwaytoys.com

Tiger Electronics
Electronic games
purchased by Hasbro in 1998
980 Woodlands Parkway
Vernon Hills, IL 60061
847-913-8100

Today's Kids
Miscellaneous
formerly Wolverine
13630 Neutron Rd.
Dallas, TX 75244
972-404-9335

Tonka
Die-cast trucks
See Hasbro

Tootsietoy / Strombecker
Vehicle toys, other
600 N. Pulaski Rd.

Chicago, IL 60624
773-638-1000

Toy Biz
Action figures
333 East 38th St.
New York, NY 10016
212-682-4700

Toy Island
Miscellaneous
100 Universal Plaza Bldg. 10
Universal City, CA 91608
818-733-7500

Toy Vault
P.O. Box 1915
London, KY 40743
606-864-8658
www.toyvault.com

Trendmasters
Action figures
611 North 10th St., Suite 555
St. Louis, MO 63101
800-771-18170

Troll and Toad
Action figures
253 Reynolds Rd.
Keavy, KY 40737-2834
606-878-2936

Ty, Inc.
Beanie Babies
P.O. Box 5377
Oakbrook, IL 60522
630-876-8000
www.ty.com

Tyco Industries
See Mattel

U.S. Games Systems
Games, playing cards
179 Ludlow St.
Stamford, CT 06902
203-353-8400

Winross
Die-cast vehicles
1965-present
Box 23860
Rochester, NY 14692
716-381-5638

Toy Clubs Directory

Editor's Note: This is a partial list of toy collector clubs. To have your club listed free in the next edition of the Toy Shop Annual, *send pertinent information to: Editor,* Toy Shop Annual, *700 E. State St., Iola, WI 54990.*

American Game Collectors Association
Specialty: Games and puzzles
Address: P.O. Box 44, Dresher, PA 19025
Newsletter: Yes
Members: 300+ members

Austin Fashion Doll Club
Specialty: Barbie dolls
Contact: Sherri Rhein
Address: 4711 Avenue F, Austin, TX 78751
Phone: 512-323-0904

Barbie and Ken Dolls Meet Wisconsin Collectors
Specialty: Barbie dolls
Contact: Katie Gorton
Address: 2545 Eastwood Ln., Brookfield, WI 53005
Members: New club

Barbie Doll Collectors Club International
Specialty: Barbie dolls
Contact: Dora Lerch
Address: P.O. Box 586, White Plains, NY 10603
Phone: 914-362-4657

Cabbage Patch Kids Collectors' Club
Specialty: Cabbage Patch Kids dolls
Address: P.O. Box 714, Cleveland, GA 30528
Phone: 706-865-2171

Chair City Barbie Club
Specialty: Barbie dolls
Address: P.O. Box 3072, Thomas-ville, NC 27361
Phone: 336-475-7110

Cracker Jack Collectors Association
Specialty: Cracker Jack and related memorabilia
Contact: Ron Toth
Address: 72 Charles St., Rochester, NH 03867-3413
Phone: 603-335-2062
Newsletter: Monthly

Hot Wheels Collectors Club
Specialty: Hot Wheels
Contact: Mattel Toys
Address: 333 Continental Blvd.,El Segundo, CA 90245-5012
Phone: 800-852-1075
Members: Company-sponsored club

Johnny Lightning News Flash
Specialty: Johnny Lightning die-cast cars
Contact: Playing Mantis Toys
Address: 3600 McGill St., Suite 300, P.O. Box 3688, South Bend, IN 46619
Phone: 219-232-0300
Newsletter: Yes
Members: 5,000 members

Larry's Traders
Specialty: Trading Cards
Contact: Larry Jackson
Address: 509 Ashland, Aurora, IL 60505
Phone: 630-851-9074
Annual Dues: 6

Looky's
Specialty: Hot Wheels
Contact: Omar
Address: 1603 N. Thurmond St., Winston-Salem, NC 27105
Members: New club

Lords of Darkness
Specialty: Fantasy, JRR Tolkien, Lord of the Rings
Contact: Paul Clement
Address: 11 Morningside Dr., Latham, NY 12110
Phone: 518-785-5099
Annual Dues: 20

M & M Collectors Club
Specialty: Promotional Items
Contact: Nancy Pinto
Address: 120 Covington Dr., Warwick, RI 02886-1936
Phone: 401-738-2277
Newsletter: M & M Happenings
Members: New club

Madison Area Die-Cast Collectors Club
Specialty: Die-Cast Vehicles
Contact: Terry F. Nadosy
Address: P.O. Box 817, Spring Green, WI 53588
Phone: 608-588-3133
Annual Dues: 12
Newsletter: Monthly
Members: Six members

Marble Collector's Society of America
Specialty: Marbles
Address: P.O. Box 222, Trumbull, CT 06611
Newsletter: Quarterly
Members: 1,000+ members

Matchbox Collectors Guild
Specialty: Matchbox Collectibles
Address: P.O. Box 10490, Glendale, AZ 85318-0490
Phone: 800-858-0102
Members: New club

Matchbox Premiere Collectors Club
Specialty: Matchbox die-cast cars
Address: P.O. Box 804, Conshohocken, PA 19428

Phone: 800-524-8697
Newsletter: Yes
Members: New company-spon-
sored club

Mazda Collectors Club
Specialty: Mazda die-cast, kits,
etc.
Contact: Werner Legrand
Address: Postbus 5, Brecht,
Belgium B2960
Phone: 323-373-4498
Newsletter: Bi-annual
Members: 20 members

**McDonald's Collectors Club,
Florida Sunshine Chapter**
Specialty: McDonald's Fast Food
Toys
Contact: Bill and Pat Poe
Address: 2220 Dominica Circle E.,
Niceville, FL 32578-4085
Phone: 850-897-4163
Annual Dues: 15
Newsletter: Sunshine Express,
monthly

Motor City Hot Wheels
Specialty: Hot Wheels
Contact: Steve Cinnamon
Address: P.O. Box 55, Belleville,
MI 48112-0055
Phone: 313-699-2170

**Nebraska, the Good Life
with Barbie**
Specialty: Barbie dolls
Contact: Laree Skeleton
Address: 735 Nye St., Fremont,
NE 68025
Phone: 402-727-6148

Pacific NW Doll Collectors Club
Specialty: Barbies, Beanies, Gene
Dolls
Contact: Kathy Anderberg
Address: 4701 225th Pl. SW,
Mountlake Terrace, WA 98043
Phone: 425-778-2442
Annual Dues: 10
Newsletter: Monthly
Members: 30+ members

PVC Collectors Club
Specialty: Cartoon and Character

Figures
Contact: Colleen Lewis
Address: 10120 Main St., Clar-
ence, NY 14031
Phone: 716-759-7451
Newsletter: Quarterly

**Smurf Collectors' Club
International**
Specialty: Smurfs
Contact: Suzann Lipschitz
Address: 24T9A Cabot Rd. W.,
Massapequa, NY 11758
Phone: 516-799-3221
Newsletter: Quarterly
Members: 3,500 members

**Star Wars Collectors Club of
Southern California**
Specialty: Star Wars
Contact: David W. Carr
Address: 20201 Burnt Tree lane,
Walnut, CA 91789-1806
Phone: 909-594-1151
Annual Dues: 5
Newsletter: Yes
Members: 30 members

**Still Bank Collectors Club
of America**
Specialty: Still banks
Contact: Larry Egelhoff
Address: 4175 Millersville Rd.,
Indianapolis, IN 46205
Phone: 317-846-7228

Suds City Hot Wheels Club
Specialty: Hot Wheels
Contact: Tim Pryal
Address: 7129 W. Moltke #3,
Milwaukee, WI 53210
Phone: 414-447-6854
Newsletter: Suds City News
Members: 125 members

The Garfield Collector
Specialty: Garfield and Related
Characters
Contact: Adrienne Warren
Address: 1032 Feather Bed Lane,
Edison, NJ 08800-1237
Phone: 732-381-7083
Annual Dues: 5
Newsletter: No
Members: 100 members

Toy Car Collectors Club
Specialty: Toy Vehicles
Contact: Peter Foss
Address: 33290 West 14 Mile Rd.
#454, West Bloomfield, MI
48322
Phone: 248-682-0272

Tucson Miniature Auto Club
Specialty: Vehicle Toys
Contact: Lou Pariseau
Address: 1111 E. Limberlost Dr.
#164, Tucson, AZ 85719-1062
Phone: 520-293-3178
Annual Dues: 10
Newsletter: Monthly
Members: 84 members

Wheels are Spinning
Specialty: Hot Wheels
Contact: Dan Hammond II
Address: 207 Kimberly Way,
Winchester, VA 22601-5579
Phone: 540-667-2430
Annual Dues: 15
Newsletter: Monthly
Members: 46 members

**Wheels of Fire Hot Wheels Club
of Arizona**
Specialty: Hot Wheels
Contact: Valerie Griffin
Address: P.O. Box 86431, Phoe-
nix, AZ 85080
Phone: 602-848-1521
Annual Dues: 20
Newsletter: Monthly

**Williams Grove Historical Steam
Engine Association**
Specialty: Antique Steam Engines
and Tractors
Contact: Show Secretary
Address: Box 509, Mechanics-
burg, PA 17055
Phone: 717-766-4001
Annual Dues: 5
Newsletter: Yes

**Windy City Collector's
Barbie Doll Club**
Specialty: Barbie dolls
Address: P.O. Box 417518,
Chicago, IL 60641

Toy Dealer Directory

Action Figures

Samantha and Bruce Christopher
P.O. Box 2810
Stony Plain, Alberta, Canada T7Z
    1Y3
403-892-2512
FAX: 403-892-2512
toybarons@globalserve.net

Figures
P.O. Box 19482
Johnston, RI 02919
401-946-5720

Bill Brady
6110 N 27th Ave.
Phoenix, AZ 85017
602-433-7546

Doug's Comics and More
1901 South 12th St., Merchants
    Square Mall
Allentown, PA 18103
610-709-0470
beanylover@aol.com

Stand Up Comics
2402 University Ave. W, Suite 411
St.Paul, MN 55114
612-646-4030

Animation Art

Mice, Ducks and Wabbits, Inc.
9045 La Fontana Blvd., Ste. B-19
Boca Raton, FL 33434
561-470-0400
FAX: 561-470-0402
wabbits@earthlink.com

Announcements / Shows

DeSalle Promotions
5106 Knollwood Ln.
Anderson, IN 46011-8729
800-392-8697

Vaughn Crispin
809 Beulah Church Rd.
Apollo, PA 15613
724-697-5510

Brimfield Associates
P.O. Box 1800
Ocean City, NJ 08226
609-926-1800

Old Mother's Cupboard
P.O. Box 4488
Mission Viejo, CA 92690
714-562-6611

Hotlanta Exposition Co.
2941 Mountain Brook Rd.
Canton, GA 30114
770-516-7212

Leading Edge Productions
P.O. Box 31388
Palm Beach Gardens, FL 33420
561-694-7982

John Carlisle
P.O. Box 1007
Lockport, NY 14095-1007

Remember When Antique &
    Collectables
6240 Chambersburg Rd.
Fayetteville, PA 17222
717-352-4762

Hedin Sportscard and Beanie Babies
    Shows
P.O. Box 1435
Framingham, MA 01701-1435
508-820-3019

Antique Toys for Sale

The Nostalgic Toy Box

204 South Main St.
Stillwater, MN 55082
651-351-7500
sclemans@msn.com

Jacqueline Henry
P.O. Box 17
Walworth, NY 14568-0017
315-986-1424
jacqueline.henry@mci2000.com

William Adorjan
P.O. Box 2494
Glenview, IL 60025
847-657-8502
iaretoys@aol.com

Steven Agin
P.O. Box 68
Delaware, NJ 07833
908-475-1796

Joe Badalucco
12908 133rd Ave.
Jamaica, NY 11420
718-845-2121

Gibson Girl Antique and Collectible
    Toys
308 S. College
Waxahachie, TX 75165
972-937-6653
FAX: 972-223-6770

Collection Connection
P.O. Box 18552
Hamilton, OH 45018
513-851-9217

Continental Hobby House
P.O. Box 193
Sheboygan, WI 53082
920-693-3371

Dan Wells Antique Toys
7008 Main St. #4
Westport, KY 40077
502-225-9925

Elegant Junque Shop
4932 E Speedway

Tucson, AZ 85712
520-881-8181
FAX: 520-881-8181

Ancient Idols Collectible Toys
223 S. Madison St.
Allentown, PA 18102
215-820-0805

Antique Toys Wanted

Mid-City Auto Antiques
409 N. Hampton Rd
DeSoto, TX 75115
972-223-6770

Gasoline Alley
6501 20th Ave. NE
Seattle, WA 98115
206-524-1606

Frank A. Najbart
2736 Bee Tree Lane
St. Louis, MO 63129-5610
314-846-2444

Auctions

Dick Brodeur
189 Arah St.
Manchester, NH 03104
603-668-4102

Ben DeVoto
756 Craig Dr.
Kirkwood, MO 63122
314-821-8588

Robert Donovan
25836 SW Rein Rd.
Sherwood, OR 97140
503-625-5464

Hake's Americana and Collectibles
P.O. Box 1444
York, PA 17405
717-848-1333

Contemporary Relics
1224 Boston Ave.
Flint, MI 48503
810-233-3202

Wanda and James Coburn
P.O. Box 71
Caldwell, ID 83606
208-459-8817

Banks, Die-Cast and Plastic

Toy Box Collectibles
525 Highland Rd. W., Suite 304
Kitchener, Ontario Canada N2H 5P4
519-570-3120
FAX: 519-895-7009
toybox@bond.net

Ken Snyder Sales
4425 Zenith Ave. So.
Minneapolis, MN 55410
612-926-5755

Bicycles, Unicycles/Tricycles, Scooters

Marky D's Bikes and Pedal Goods
7047 Springridge Rd.
West Bloomfield, MI 48322
248-398-0660
FAX: 248-398-2581
toyman047@aol.com

Books

Blystone's
2132 Delaware Ave., Dept. TS
Pittsburgh, PA 15218
412-371-3511

Buttons

Stormy's Pinbacks and Collectibles
4867 Greenleaf Rd.
Sarasota, FL 34233
941-921-5926

Cars for Sale

Zurko's Midwest Promotions
211 W. Green Bay St.
Shawano, WI 54166
715-526-9769

FAX: 715-524-5675

Mike Silver
509 Danielle Ct.
Roseville, CA 95747
916-782-4800

Bob Barnes
43 Via Alicia
Santa Barbara, CA 93108
805-962-9559

Superior Diecast
P.O. Box 124
Little Lake, MI 49833
906-346-9421
FAX: 906-346-3729

Die-Cast Delirium
P.O. Box 4451
Culver City, CA 90231
diecastdel@aol.com

Supercar Collectibles
7311 75th Circle North
Brooklyn Park, MN 55428
612-425-6020
FAX: 612-425-6020
jimthoren@wavefront.com

Dixie Diecast
10 Gobar Landing Rd.
Statesboro, GA 30461
888-842-5994
FAX: 912-823-3326

The Miniature Transport Shop
P.O. Box 520
Forestville, New South Wales,
    Australia 2087
messmod@mph.com.au

Liverpool Motor Works
RD #1, Box 19B
Liverpool, PA 17045
717-444-3773
FAX: 717-444-2457
lmwjerry@aol.com

Champion Toys & Automobilia
RR1, Box 1858
Kennebunkport, ME 04046
207-985-2292

Louis Panuse
8113 12th Ave.SW
Seattle, WA 98106
206-764-7131

Cars for Sale (Slot Cars)

Futuretronics
2055 North Ridge Rd
Lorain, OH 44055
440-277-8004

Cars Wanted

The Little Car Shop
Rt. 2, Box 151E
Kilgore, TX 75662
903-984-1806

Catalogs, Collectible

Thomas De Stefano
7629 W. Norridge St.
Harwood Heights, IL 60656-3348

Cereal Boxes & Premiums

Roland Coover
1537 E. Strasburg Rd.
West Chester, PA 19380
610-692-3112
FAX: 610-738-9108

Chris Serratore
218 N. Main St.
Ambler, PA 19022
215-628-3213
FAX: 215-643-4558
funface@bellatlantic.net

Character Toys for Sale

Annie M's Collectibles
P.O. Box 3, Dept. TSAD
East Moline, IL 61244-0003

309-755-1052
anniemoz@aol.com

Harry Greenfield: DBA Then and
    Again
P.O. Box 344
Marlborough, CT 06447-0344
860-295-9896
860-295-0839

Lurch and DeeAnn
c/o State net, 2101 K St.
Sacramento, CA 95816
916-487-3251
lurch-deeann@iname.com

Adrienne Warren
1032 Feather Bed Lane
Edison, NJ 08820-1237
732-381-7083

Heathside Collectibles
P.O. Box 1416
Biddeford, ME 04005
hcollect@lamere.net

Casey's Collectible Corner
HCR Box 31, Rte. 30
North Blenheim, NY 12131
607-588-6464

TV Toyland
223 Wall St.
Huntington, NY 11743
516-385-1306
FAX: 516-385-1307
itsonlyrocknroll@erols.com

Buffalo Road Hobby Imports
10120 Main St.
Clarence, NY 14031
716-759-7451
FAX: 716-759-7452
bripvc@toyline.com

Thomas W. Bowman
529 8th St. SE
Minneapolis, MN 55414
651-274-6238

Global Pop Culture
P.O. Box 11397
Torrance, CA 90510-1397
310-512-6000
FAX: 310-281-6000
globalpop@aol.com

M & M Collectibles
120 Covington Dr.
Warwick, RI 02886
401-738-2277

Queen's Collection, The
P.O. Box 235
Clifton, NJ 07011
973-772-8548

Melanie Kiviat
P.O. Box 57754
Sherman Oaks, CA 91413
818-788-0774

Anne Michaels, Toyologist
5917 Cerritos Ave.
Cypress, CA 90630
714-821-7990

Character Toys Wanted

Clif Smith, "Car 54 Where are You?"
P.O. Box 110413
Nashville, TN 37222
615-297-2978
csmiths@worldnet.att.net

Comics for Sale

Good Time Charlies
114 W. Knox
Ennis, TX 75119
972-875-9737

Best Comics
252-02 Northern Blvd.
Little Neck, NY 11362
718-279-2099

Diamond International Galleries
1966 Greenspring Dr. Suite 300
Timonium, MD 21093
410-560-5810

Avalon Comics
P.O. Box 234
Boston, MA 02123
617-262-5544

Dinky Toys

Scot Marechaux
6234 Glosser Rd
Belmont, NY

716-268-5582
FAX: 716-268-5533

Doll Furniture

John Madore
13489 Route 30
Irwin, PA 15642
724-864-1571

Dolls for Sale, Antique

Raggedy's and Teddy's Co.
6337 Nightwind Circle
Orlando, FL 32818-8834
407-884-5483
FAX: 407-884-5483
raggedyman@aol.com

Dolls for Sale, Barbie

Marl & B
10301 Braden Run
Bradenton, FL 34202
941-751-6275

Doll Attic, The
2491 Regal Dr.
Union City, CA 94587
510-489-0221

Dolls for Sale, Other

Ertl Toys

Tom-C-Toys
8095 St. Rt. 305
Garrettsville, OH 44231-0185
330-527-7601
FAX: 330-527-9052
totoys@apk.net

Farm Tractors

Melvin E. and Melody L. Enck
501 Windy Hill Rd., Lot 157
Shermans Dale, PA 17090
717-439-6621

Gene's Toys & Collectibles
207 East Welch, P.O. Box 327

Crescent, IA 51527
712-545-9306
FAX: 712-545-9306
genestoys@uswest.net

Fast Food Collectibles

Nancy Pinto
120 Covington Dr.
Warwick, RI 02886-1936
401-738-2277
uki845@aol.com

Bill and Pat Poe
220 Dominica Circle E.
Niceville, FL 32578-4085
850-897-4163
FAX: 850-897-2606
McPoes@aol.com

Fire Engines Wanted

Collectibles and Philatelic Services International
Suite 333, 236 Hyperdome, Bryants Rd. and Pacific Hwy.
Loganholme, Australia 4129
valerieandrewhenry@bigpond.com

G.I. Joe

Cotswold Collectibles
P.O. Box 716TS
Freeland, WA 98249
360-331-5331

Boom Gallery
146 N. Northwest Highway
Park Ridge, IL 60068
847-823-2666

Joe Depot, The
P.O. Box 228
Kulpsville, PA 19443
215-721-9749

Games, Board & Boxed

For Keep's Sake
4224 Eagle Head Dr.
Columbus, OH 43230
614-475-1101
info@rareinvest.com

Bill Smith
56 Locust St.
Douglas, MA 01516
508-476-2015

Guns, Air and Pop

Jerry Lorentz Collectables
43769 Vista Rd.
Isle, MN 56342

Hot Wheels

Cape Coral Cards
1311 Del Prado Blvd.
Cape Coral, FL 33990
941-772-8222

Adkins Collectibles
422 East Oak St.
Oak Creek, WI 53154-1121

414-761-1020
FAX: 414-761-1088

Dale A. Austin
15 Russell St.
Bangor, ME 04401-4533
207-942-4527

Bob Goforth
4061 E. Castro Valley Blvd. Suite 224
Castro Valley, CA 94552
510-889-6676

Frank's Comics and Cards
2678 S.W. 87 Ave.
Miami, FL 33165
305-226-5072

Griffin Collectibles
6977 W. San Miguel Ave.
Glendale, AZ 85303
602-848-1521

Tim Pryal
7129 W. Moltke #3
Milwaukee, WI 53210
414-447-0854

Matchbox

Roger Behm
122 Shippen Dr.
Coraopolis, PA 15108
412-264-7770

Dave Phillipo
582 Lincoln St.
Marlboro, MA 01752
508-481-6610

Military Soldiers

Relic Golden Amusements
P.O. Box 572
Hackensack, NJ 07603
201-342-6475
FAX: 201-342-5954

Miscellaneous Toys for Sale

All-Star Celebrity Collectibles
5637 Keokuk Ave.
Woodland Hills, CA 91367
818-884-2969

White Salmon Collectibles
3120 SW Scholls Ferry Rd..
Portland, OR 97221
503-292-0536

Whiz Bang Collectibles
P.O. Box 300546
Fern Park, FL 32730
407-260-8869

Funk & Junk
Alexandria, VA 22306
703-836-0749
junkmail@funkandjunk.com

Wonderland Toys
236 Pepper Ridge Circle
Antioch, TN 37013
615-781-2190
FAX: 615-781-2190

Yankee Peddler Antiques and Collectibles
P.O. Box 205
Lothian, MD 20711
410-741-9080
FAX: 410-741-9123

John's Collectible Toys
57 Bay View Dr.
Shrewsbury, MA 01545
508-852-0005

Serious Toyz
P.O. Box 575
Port Washington, NY 11050
516-944-8941
serioustoyz@pipeline.com

Dennis Allen
740 N. Munsterman St.
Appleton, MN 56208
302-289-2493

Planet Toyz
6720 Hwy. 90 East #2
Morgan City, LA 70380
504-385-5771
toyman@iamerica.net

Lou Englehart
187 Three Bridge Rd.
Monroeville, NJ 08343
609-358-8625

Sally's Antiques and Collectibles
196 Front St.
Owego, NY 13827
607-687-4111

Cynthia L. Case
11324 Knollridge Ct. #3
Indianapolis, IN 46229-3711
317-254-2735

Jenkinson's Tiques N' Toys
8305 N. Highland
Kansas City, MO 64118
816-468-6676

Halloween Outlet
246 Park Ave.
Worcester, MA 01609
508-798-9957

Sonny's Toys & Collectibles
6 Hendel Dr., Rout 27
Old Mystic, CT 06372
860-442-4572

Jeff & Bob's Fun Stuff
P.O. Box 370332
Reseda, CA 91335
818-831-4157

Toy Go Round
1850 Berryville Pike
Winchester, VA 22603
540-662-8818
FAX: 540-662-1807
toygoround@hotmail.com

Cosmic Toys
93 Oak St.
Wyandotte, MI 48192
734-282-9598

California Collectables
1060 E. 11th St.
Oakland, CA 94606
510-653-1310

Bill Campbell
1221 Littlebrook Ln.
Birmingham, AL 35235
205-853-8227

Collectorholics
15006 Fuller
Grandview, MO 64030
816-322-0906

Barry Goodman
P.O. Box 218
Woodbury, NY 11797
516-338-2701

Tom Lastrapes
P.O. Box 2444
Pinellas Park, FL 33780
727-545-2586

M & J Variety
932 E. Boulevard
Alpha, NJ 08865
908-213-9099

Mad About Toys
1525 Aviation Blvd. #109
Redondo Beach, CA 90278
310-318-1829

Cyco for Toys
P.O. Box 1
Atlantic Beach, NY 11509
516-623-7317

Darrow's Fun Antiques
1101 First Ave.
New York, New York 10021
212-838-0730

Quest for Toys
P.O. Box 3519
Alliance, OH 44601
330-829-5946

Splash Page Comics
P.O. Box 158
Kirksville, MO 63501
660-665-7623

B & D Comics & Toys
8A Union St.
Montgomery, NY 12549
914-457-1237

Michael Melito
29A Foodmart Rd.
Boston, MA 02118
781-662-2189

Cricket Hill Toys
1175 County Rt. 1

Oswego, NY 13126
315-343-6715

Another World Collectibles
P.O. Box 47
Gays, IL 61928
217-752-6214

Steve's Lost Land of Toys
3572 Turner Ct.
Fremont, CA 94536
510-795-0598

Amok Time Toys
2941 Hempstead Turnpike
Levittown, NY 11756
516-520-0975

Collectible Madness
483 Federal Rd.
Brookfield, CT 06804
203-740-9472

Miscellaneous Toys Wanted

Alan Rosen
70 Chestnut Ridge Rd., Suite I
Montvale, NJ 07645

Model Kits

Shadowland
1637 Hanover Ave.
Allentown, PA 18103
610-437-0189

Greg Long
215 Williamsburg
Montgomery, IL 60338
630-801-0110

Motorcycles

Jeff Hubbard
2900 91st St.
Sturtevant, WI 53177
414-886-0477

Movie and TV Show Toys

James Koval
16907 Bougainvilla Ln.
Friendswood, TX 77546-3417
281-648-4787
FAX: 281-648-4788

Marshall Turner
270 Ku'ulep Rd.
Kailua, HI 96734
808-263-0072

Jan Polla
16-64 155 St.
Whitestone, NY 11357-3233
718-746-0911
FAX: 718-746-0911

Paper Dolls

Janice Hanks
116 School St.
Littleton, NH 03561
603-444-5134

PEZ

Steve Glew
5611 Lehman Rd.
Dewitt, MI 48820
517-669-5931

Nouveau Technologies
P.O. Box 201021
Arlington, TX 76006
817-801-8008
numbervi@aol.com

Pedal Cars

MID-CITY ANTIQUES COLLECTIBLES

BUYING
Pedal Cars
Boats
Planes
Tractors
Antiques

SELLING
Cast-Iron   Ertl Banks
Pressed Steel   Oil-Gas
Toys   Collectibles

Merle J. Vondrasek
409 N. Hampton
DeSoto, TX 75115
972-223-6770
972-223-4087

Services / Insurance

Collectibles Insurance Company
P.O. Box 1200
Westminster, MD 21158

888-837-9537

Space Toys

Image Anime Co. Ltd.
103 W. 30th St.
New York, NY 10001
212-631-0966

Cincinnati Sci-Fi
7865 Intl. Dayton Rd.
Westchester, OH 45069
513-759-9533

Outer Limits
433 Piaget Ave.
Clifton, NJ 07013
973-340-9393

Tom's Sci-Fi Shop
P.O. Box 56116
Harwood Heights, IL 60656-0116
708-867-8291

Canstar
1402 Pine Ave., Suite 105
Niagara Falls, NY 14301
905-832-5350

Brian Semling
P.O. Box 95, W730 Hwy. 35
Fountain City, WI 54629
608-687-7572

Sports Related for Sale

Phil's Line-ups and Figures
58 Sunrise Dr.
Lynbrook, NY 11563
516-733-0852
FAX: 516-887-5880
pmblock@banet.net

Toy Repairs

Greg Allred
7012 Trading Post Lane
Las Vegas, NV 89128
702-255-4033

Joseph E. Degrella
P.O. Box 281
Buckner, KY 40010
502-241-4509

Trading Cards

Larry's Sportscards
509 Ashland
Aurora, IL 60505
630-851-9074

Barrington Square
P.O. Box 310-E
West Dundee, IL 60018
847-426-2020

Trucks for Sale

Jim "Mr. Winross" Lord
1122 Main St.
Evanston, IL 60202-1649
847-864-9243
FAX: 847-864-5978
winross@sprintmail.com

Tom Snook
478 Sandy Way
Benicia, CA 94510-2621
510-787-3283
FAX: 510-787-3293

Ronald Burmeister
P.O. Box 532
Bayport, NY 11705-0532
516-758-3455
FAX: 516-758-3979

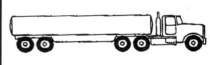

Won Collectibles
61-2C Taurus Dr.
Hillsborough, NJ 08876
908-281-0451
FAX: 732-752-6754

Norman Humiston
1117 Rt. 29
Schuylerville, NY 12871
518-695-3000

Vehicles, Die-Cast and Plastic

Morrill's Miniature Motors
2901 N. 10th #F
McAllen, TX 78501
956-687-4342
FAX: 956-687-8616

Quality Steins & Collectibles
Box 762
Bowling Green, OH 43402
419-353-6847

Asheville Diecast
1434 Brevard Rd.
Asheville, NC 28806
828-667-9690

Werner Legrand
Postbus 5
Brecht, Belgium B2960
323-313-4498

Carolina Hobby Expo
3452 O'Dell School Rd.
Concord, NC 28027
704-786-8373

Blast from the Past
2105 1st St.
Ann Arbor, MI 48104
313-996-2765

Frank and Ruth Abramson
7146 Altama Rd.
Jacksonville, FL 32216-9108
904-724-0748
truck2car@aol.com

Bruce Johnson Toy Talk
702 Steeplechase Rd.
Landisville, PA 17538
717-898-2932

Budget-Minder Collectibles
701 E. Bay St., Box 1017
Charleston, SC 29403

Toys for Collectors
P.O. Box 1406
Attleboro Falls, MA 02763
508-695-0588

G. Cleveland
12 High St.
Rickmansworth, Herts, Great Britain WD31ER

British Triumph and Metropolitan
9957 Frederick Rd.
Ellicott City, MD 21042-3647
410-750-2352
pete_groh@yahoo.com

Coll. Toy and Model Shop
1 Fitzroy Ave.
Kent, England
484-329-1905

Vehicles: Die-Cast and Plastic

Last Chance Garage
5550 Cornhusker Hwy.
Lincoln, NE 68504
877-464-1171
FAX: 402-464-1171
jeff@inetnebr.com

Loren Miller Promos
6480 S. Lone Elder Rd.
Aurora, OR 97002
503-678-1335

Diecast Miniatures
255 Jones St., P.O. Box 58
Amston, CT 06231-0058
860-228-3249

Iron Horse Antiques and Collectibles
100 Hurlbut St.
Winsted, CT 06098
860-379-5011
FAX: 860-379-5011
jlauretti@snet.net

Western Toys

Black Bart's Trading Post
320 S. National
Springfield, MO 65802
417-863-6168
FAX: 417-863-0345
ddenton823@aol.com

Quarterback Sneak
P.O. Box 1091
Eau Claire, WI 54702
715-334-7934

Mike's Western
440 E. Pearl St.
Geneseo, IL 61254
309-944-4423

Wind-Up and Battery Toys

Robots and Wind Me Up Toys
713 Hampton St.
Tipp City, OH 45371
937-667-7587
FAX: 937-667-5687
theinl@juno.com

It's Toy Time
19360-81st Pl. North
Maple Grove, MN 55311
612-420-7383
FAX: 612-425-7032

1999 Toy Show Calendar

ALABAMA

Jan 30-31 1999 AL, Birmingham. Greenberg's Great Train & Collectible Toy Show. State Fairgrounds, 2331 Bessemer Rd. SH: Sat. 11am-5pm, Sun. 11am-4pm, A: $5., $2. ages 6-12, under 6 free. Greenberg Shows, Nan Turfle, 7566 Main St., Sykesville, MD 21784. PH: 410-795-7447.

ARIZONA

Mar 7 1999 AZ, Tucson. 11th Annual Collectible Toy Show. Marketplace USA, 3750 E. Irvington Rd. SH: 9am-3pm, A: $2., under 13 free with adult. Lou Pariseau, PH: 520-293-3178 or Jerry Rettig, PH: 520-322-9881.

Apr 17 1999 AZ, Phoenix. Doll, Bear & Toy Show. Shrine Auditorium, 552 N. 40th St. SH: 10am-4pm, A: $4., under 12 free. Fran O., PH: 602-994-5361 or Fran J., PH: 602-947-3981.

ARKANSAS

May 15 1999 AR, Eureka Springs. 18th Annual Doll & Toy Show. Inn of the Ozarks Convention Center, Rt. 62 West. SH: 10am-4pm, T: 80. Marilyn Bromstad, 279 County Road 226, Eureka Springs, AR 72631. PH: 501-253-2244 eves.

Sep 11 1999 AR, Eureka Springs. 12th Annual Toys, Dolls & Trains Show. Convention Ctr., Inn of the Ozarks, Best Western Motel, Hwy. 62. SH: 9am-4pm, T: 80. George Bromstad, 279 County Road 226, Eureka Springs, AR 72631. PH: 501-253-2244 eves.

CALIFORNIA

Jan 2-3 1999 CA, San Rafael. Greenberg's Great Train & Collectible Toy Show. Marin Ctr., Avenue of the Flags. SH: Sat. 11am-5pm, Sun. 11am-4pm, A: $5., $2. ages 6-12, under 6 free. Greenberg Shows, Nan Turfle, 7566 Main St., Sykesville, MD 21784. PH: 410-795-7447.

Jan 3 1999 CA, Fresno. Beanie Babies & Sports Card Show. Ramada Inn, 324 E. Shaw Ave. & 41 Fwy. SH: 10am-4pm, T: 50-8'. Jim or John, PH: 209-322-0407 or 909-944-7628.

Jan 3 1999 CA, Petaluma. Toy Show. Veterans Bldg., 1094 Petaluma Blvd. S. SH: 10am-3pm, T: 70. Linda Schroeder or Mel Pace, 304 Petaluma Blvd. N., Petaluma, CA 94954. PH: 707-776-4566.

Jan 3 1999 CA, Stanton-Orange County. Toy, Beanie Babies, Barbie & Model Train Show. Katella Center, 7520 Katella Ave. SH: 9am-2:30pm, A: $2., 10 & under free. Ron, PH: 909-902-0217.

Jan 9-10 1999 CA, San Diego. Teddy Bear, Doll & Antique Toy Show. Scottish Rite Ctr., 1895 Camino Del Rio S. SH: Sat. 10:30am-4pm, Sun. 10am-3pm, T: 150. Linda Mullins, PO Box 2327, Carlsbad, CA 92018. PH: 760-434-7444 or FAX: 760-434-0154.

Jan 10 1999 CA, San Mateo. Toy Show. County Expo Center, 2495 S. Delaware St. SH: 10am-3pm. Dugger-Hill Promotions, PH: 530-888-0291 or 408-476-8612.

Jan 17 1999 CA, Fresno. Beanie Babies & Sports Card Show. Ramada Inn, 324 E. Shaw Ave. & 41 Fwy. SH: 10am-4pm, T: 50-8'. Jim or John, PH: 209-322-0407 or 909-944-7628.

Jan 23 1999 CA, Santa Rosa. Antique to Modern Dolls, Teddy Bears & Beanies Show. Sonoma Cty. Fairgrounds, 1350 Bennett Valley Rd., Hwy. 101 to Hwy. 12 E. SH: 10am-4pm, A: $5., $2. 12 & under, 5 & under free. Golden Gate Shows, Fern Loiacono, PO Box 1208, Ross, CA 94957. PH: 415-662-9500 or FAX: 415-662-9600.

Feb 6 1999 CA, San Jose. Antique to Modern Dolls, Teddy Bears & Beanies Show. Santa Clara Cty. Fairgrounds, 344 Tully Rd. to Tully Rd., Exit W. SH: 10am-4pm, A: $5., $2. 12 & under, 5 & under free. Golden Gate Shows, Fern Loiacono, PO Box 1208, Ross, CA 94957. PH: 415-662-9500 or FAX: 415-662-9600.

Feb 13 1999 CA, San Mateo. NNL Western Nationals Model Car Convention. Expo Center. NNL Western Nationals, PO Box 149, Santa Clara, CA 95050. PH: 510-582-9234.

Feb 14 1999 CA, San Mateo. West Coast Model Expo Swap Meet. .Expo Center. West Coast Model Expo, PO Box 149, Santa Clara, CA 95050. PH: 510-582-9234.

Feb 20 1999 CA, San Francisco. Antique to Modern Dolls, Teddy Bears & Beanies Show. Cow Palace, 2600 Geneva Ave., Daly City, Hwy. 101 or 280 to Cow Palace Exit to Geneva Ave. SH: 10am-4pm, A: $5., $2. 12 & under, 5 & under free. Golden Gate Shows, Fern Loiacono, PO Box 1208, Ross, CA 94957. PH: 415-662-9500 or FAX: 415-662-9600.

Mar 7 1999 CA, Sacramento. Antique Toy Show. Scottish Rite Temple, 6151 H St. SH: 10am-3pm, T: 100. Wayne Gateley Productions, PO Box 221052, Sacramento, CA 95822. PH: 916-448-2655.

Mar 20 1999 CA, Marin County. Antique to Modern Dolls, Teddy Bears & Beanies Show. Civic Ctr., Exhibit Hall, Ave. of the Flags, San Rafael, Hwy. 101 to N. San Pedro Rd. SH: 10am-4pm, A: $5., $2. 12 & under, 5 & under free. Golden Gate Shows, Fern Loiacono, PO Box 1208, Ross, CA 94957. PH: 415-662-9500 or FAX: 415-662-9600.

Mar 28 1999 CA, San Mateo. Toy Show. County Expo Center, 2495 S. Delaware St. SH: 10am-3pm. Dugger-Hill Promotions, PH: 530-888-0291 or 408-476-8612.

Apr 30-May 2 1999 CA, Clovis. Beanie Baby, Sports Card & Collector Show. Sierra Vista Mall. T: 8'. John, PH: 209-322-0407 or Jim, PH: 909-944-7628.

May 1 1999 CA, Monterey. Antique to Modern Dolls, Teddy Bears & Beanies Show. Cty. Fairgrounds, 2004 Fairground Rd., Hwy. 1 to Casa Verde Exit. SH: 10am-4pm, A: $5., $2. 12 & under, 5 & under free. Golden Gate Shows, Fern Loiacono, PO Box 1208, Ross, CA 94957. PH: 415-662-9500 or FAX: 415-662-9600.

May 2 1999 CA, Hayward. Antique Toy Show. Centennial Hall, 22292 Foothill Blvd. SH: 10am-3pm, T: 200. Wayne Gateley Productions, PO Box 221052, Sacramento, CA 95822. PH: 916-448-2655.

May 22-23 1999 CA, Pasadena. World Toy & Collectible Expo. Pasadena Ctr., 300 E. Green St. SH: 11am-7pm, T: 100. Creation Entertainment, Galleria Tower, 100 W. Broadway, 12th Fl., Glendale, CA 91210. PH: 818-409-0960.

May 22-23 1999 CA, Pasadena. Fangoria's Weekend of Horrors. Pasadena Ctr. Creation Entertainment, Galleria Tower, 100 W. Broadway, 12th Fl., Glendale, CA 91210. PH: 818-409-0960.

Jun 5 1999 CA, San Jose. Antique to Modern Dolls, Teddy Bears & Beanies Show. Santa Clara Cty. Fairgrounds, 344 Tully Rd., Hwy. 101 to Tully Rd., Exit W. SH: 10am-4pm, A: $5., $2. 12 & under, 5 & under free. Golden Gate Shows, Fern Loiacono, PO Box 1208, Ross, CA 94957. PH: 415-662-9500 or FAX: 415-662-9600.

Jun 12 1999 CA, Santa Rosa. Antique to Modern Dolls, Teddy Bears & Beanies Show. Sonoma Cty. Fairgrounds, 1350 Bennett Valley Rd., Hwy. 101 to Hwy. 12 E. SH: 10am-4pm, A: $5., $2. 12 & under, 5 & under free. Golden Gate Shows, Fern Loiacono, PO Box 1208, Ross, CA 94957. PH: 415-662-9500 or FAX: 415-662-9600.

Aug 14 1999 CA, Marin County. Antique to Modern Dolls, Teddy Bears & Beanies Show. Civic Ctr., Exhibit Hall, Ave. of the Flags, San Rafael, Hwy. 101 to N. San Pedro Rd. SH: 10am-4pm, A: $5., $2. 12 & under, 5 & under free. Golden Gate Shows, Fern Loiacono, PO Box 1208, Ross, CA 94957. PH: 415-662-9500 or FAX: 415-662-9600.

Sep 11 1999 CA, San Jose. Antique to Modern Dolls, Teddy Bears & Beanies Show. Santa Clara Cty. Fairgrounds, 344 Tully Rd., Hwy. 101 to Tully Rd., Exit W. SH: 10am-4pm, A: $5., $2. 12 & under, 5 & under free. Golden Gate Shows, Fern Loiacono, PO Box 1208, Ross, CA 94957. PH: 415-662-9500 or FAX: 415-662-9600.

Nov 7 1999 CA, Hayward. Antique Toy Show. Centennial Hall, 22292 Foothill Blvd. SH: 10am-3pm, T: 200. Wayne Gateley Productions, PO Box 221052, Sacramento, CA 95822. PH: 916-448-2655.

Nov 14 1999 CA, Sacramento. Antique Toy Show. Scottish Rite Temple, 6151 H St. SH: 10am-3pm, T: 100. Wayne Gateley Productions, PO Box 221052, Sacramento, CA 95822. PH: 916-448-2655.

CONNECTICUT

Jan 3 1999 CT, Waterbury. A&M Collectible Toy Show. Four Points Sheraton, 3580 E. Main St., I-84E, Exit 25A, I-84W, Exit 26. SH: 9am-3pm, T: 100-8'. Bernie, PH: 860-274-9592.

Jan 10 1999 CT, Waterbury. A Model Railroad Show. Four Points Hotel, 3580 E. Main St., Rt. 84 E. Exit 25A, Rt. 84 W. Exit 26. SH: 9am-3pm. LLC, PO Box 2415, Shelton, CT 06484. PH: 203-926-1327.

Jan 19 1999 CT, New Haven. Train, Toy & Collectibles Show. Annex Y.M.A. Hall, 554 Woodward Ave. A: $3., children free. Frank Schiavone, 20 Boston Ave., New Haven, CT 06512. PH: 203-467-3133.

Feb 7 1999 CT, Norwich. SE CT Toy & Collectible Show. Ramada Hotel, 395 to Exit 80 or 80 W. SH: 9am-3pm, T: 100-8'. Jim Arpin, 104 Kinsman Rd., Lisbon, CT 06351. PH: 860-822-8514.

Feb 21 1999 CT, Waterbury. A&M Collectible Toy Show. Four Points Sheraton, 3580 E. Main St., I-84E, Exit 25A, I-84W, Exit 26. SH: 9am-3pm, T: 100-8'. Bernie, PH: 860-274-9592.

Mar 28 1999 CT, Norwich. SE CT Toy & Collectible Show. Ramada Hotel, 395 to Exit 80 or 80 W. SH: 9am-3pm, T: 100-8'. Jim Arpin, 104 Kinsman Rd., Lisbon, CT 06351. PH: 860-822-8514.

DELAWARE

Jan 10 1999 DE, Hockessin. Prime Mover Train Show. Memorial Hall, Rt. 41. SH: 9am-2pm, T: 75-8'. Tom Marinelli, 117 Westview Dr., Lincoln Univ., PA 19352. PH: 610-255-4785.

Feb 7 1999 DE, New Castle. Automotive Toys & Models NASCAR Collectibles Show. Nur Temple, Rt. 13 N. (Rt. 40 Split). SH: 9am-3pm, T: 90, A: $3., under 12 free. High Speed Promotions, Steven Rosenzweig, 349 Windsor Dr., Cherry Hill, NJ 08002. PH: 609-667-6808.

FLORIDA

Jan 2-3 1999 FL, Daytona Beach. 21st Annual Doll Show. Ramada Inn Resort Oceanfront, 2700 N. Atlantic Ave. SH: 10am-4pm. Barbara Gatfield, 487 Hopi Ct., Port Orange, FL 32127. PH: 904-767-0008.

Jan 2-3 1999 FL, Orlando. Collectibles & Toy Expo. FL Fairgrounds, 4603 W. Colonial Dr. SH: Sat. & Sun. 10am-5pm, A: $5. Classic Shows, Chip Nofal, PO Box 1507, St. Augustine, FL 32085. PH/FAX 904-928-0666.

Jan 2-3 1999 FL, Pompano Beach. Greenberg's Great Train & Collectible Toy Show. Broward Comm. College, The Omni, 1000 Coconut Creek Blvd. SH: Sat. 11am-5pm, Sun. 11am-4pm, A: $5., $2. ages 6-12, under 6 free. Greenberg Shows, Nan Turfle, 7566 Main St., Sykesville, MD 21784. PH: 410-795-7447.

Jan 9-10 1999 FL, Jacksonville. Greenberg's Great Train & Collectible Toy Show. Fairgrounds, across from Coliseum. SH: Sat. 11am-5pm, Sun. 11am-4pm, A: $5., $2. ages 6-12, under 6 free. Greenberg Shows, Nan Turfle, 7566 Main St., Sykesville, MD 21784. PH: 410-795-7447.

Jan 10 1999 FL, Ft. Lauderdale. Barbie Goes To FL Lauderdale. Airport Hilton, 1870 Griffen Rd. SH: 10am-3pm, A: $5., $2. under 12. Marl, PH: 941-751-6275 or Joe, PH: 213-953-6490.

Jan 16-17 1999 FL, Orlando. Florida Extravaganza FX-99. Orange Cty. Convention Ctr., 9800 International Dr. SH: Sat. 10am-6pm, Sun. 10am-4pm, T: 1,000. Laura or Bruce Zalkin, PH: 941-343-0094.

Jan 23-24 1999 FL, Tampa. Greenberg's Great Train & Collectible Toy Show. State Fairgrounds, 4800 US Hwy. 301 N. SH: Sat. 11am-5pm, Sun. 11am-4pm, A: $5., $2. ages 6-12, under 6 free. Greenberg Shows, Nan Turfle, 7566 Main St., Sykesville, MD 21784. PH: 410-795-7447.

Jan 23 1999 FL, Ft. Lauderdale. South Florida Collectible Toy & Doll Super Show. Elks Exhibition Center, 700 NE 10th St., Pompano Beach. SH: 10am-5pm, T: 150-8'. Mark Leinberger, PH: 561-694-7982 or Jon Jacobus, PH: 954-772-1420.

Jan 30-31 1999 FL, Jacksonville. Extravaganza. FL Fairgrounds, 510 Fairgrounds Place. SH: Sat. & Sun. 10am-5pm, A: $5. Classic Shows, Chip Nofal, PO Box 1507, St. Augustine, FL 32085. PH/FAX 904-928-0666.

Jan 30 1999 FL, Orlando. 2nd Annual FL Raggedy Ann, Doll & Teddy Bear Convention. Bahia Shrine Auditorium, 2300 Pembrook Dr. SH: 9am-4pm. Larry Vaughan, 6337 Nightwind Cir., Orlando, FL 32818. PH: 407-884-5483.

Jan 30 1999 FL, Sarasota. Train Show. Armory, 2890 Ringling Blvd. SH: 10am-4pm, T: 100. Steven Harris, 2581 Countryside Blvd. #203, Clearwater, FL 33761. PH: 813-791-1621.

Jan 30-31 1999 FL, Palmetto. Doll, Bear & Beanies Show. Manatee Civic Ctr., US Hwy. 41. SH: Sat. 9am-4pm, Sun. 11am-4pm. Neva Winkel, PH: 941-722-6675 or Lanell Rowland, PH: 941-792-8487.

Jan 30 1999 FL, Pompano Beach. Doll, Bear & Toy Show. Recreation Ctr., 1801 NE 6th St. SH: 10am-4pm. Susan, PH: 305-652-8237.

Jan 31 1999 FL, Sarasota. Toy Show. Armory, 2890 Ringling Blvd. SH: 10am-4pm, T: 100. Steven Harris, 2581 Countryside Blvd. #203, Clearwater, FL 33761. PH: 813-791-1621.

Feb 7 1999 FL, W. Palm Beach. 5th Annual FL Toy Soldier & Action Figure Show. Airport Holiday Inn, 1301 Belvedere Rd. at I-95. Frank Burns, 715 SW 15th St., Boynton Beach, FL 33426. PH: 561-732-7295 or FAX: 561-734-3842.

Feb 7 1999 FL, Miami. Antique Toy, Doll & Collectibles Show. Airport Marriott Hotel, 1201 NW LeJeune Rd. (LeJeune Rd. S. Exit, SR 836). SH: 10am-4pm. PH: 305-446-4488 or 669-2550.

Feb 13-14 1999 FL, Auburndale-Lakeland. Central FL Toy & Collectible Expo. Polk Cty. Fairgrounds, Auburndale Speedway, Hwy. 542. SH: 8am-4pm, T: 850. Doug Malcolm, 1420 N. Galloway Rd., Lakeland, FL 33810. PH: 941-686-8320 or Bill & Pat Poe, PH: 850-897-4163 or FAX: 850-897-2606.

Feb 13-14 1999 FL, Auburndale. FL Sunshine Chapter-McDonald's Collectible Club Show. Motels in Winter Haven & Lakeland, Hwy. 542, Polk County Fairgrounds. SH: 8am-4pm, T: 900. Bill & Pat Poe, 220 Dominica Circle E., Niceville, FL 32578. PH: 850-897-4163 or FAX: 850-897-2606.

Mar 6-7 1999 FL, Orlando. Collectible Toy & Doll Super Show. Bahia Shrine Auditorium, 2300 Pembrook Dr. SH: Sat. 10am-5pm, Sun. 10am-4pm, T: 200-8'. Mark Leinberger, PH: 561-694-7982 or Jon Jacobus, PH: 954-772-1420.

Mar 13-14 1999 FL, Ft. Lauderdale. Tate's Comics Presents Toy, Comic, Model, Doll, Non-Sport Cards Beanie Baby Extravaganza. War Memorial Auditorium, 800 NE 8th St. SH: Sat. 10am-5pm, Sun. 10am-5pm, T: 350. PH: 954-748-0181.

Mar 13 1999 FL, Jacksonville. River City Antique Doll &

Collectable Show. Morocco Temple, St. Johns Bluff Rd. SH: 10am-4pm, A: $4., $1., under 10. M. Hill, 13030 Bent Pine Ct. E., Jacksonville, FL 32246. PH: 904-221-1235.

Mar 20 1999 FL, Kissimmee. 5th Semi-Annual Doll & Bear Show. Agricultural Ctr., 1901 E. Irlo Bronson Hwy. (Hwy. 192). SH: 10am-4pm, A: $4., 12 & under free with adult. Steve Schroeder, 3100 Harvest Ln., Kissimmee, FL 34744. PH: 407-957-6392 or FAX: 407-957-9427.

Apr 10 1999 FL, Ft. Lauderdale. South Florida Collectible Toy & Doll Super Show. Elks Exhibition Center, 700 NE 10th St., Pompano Beach. SH: 10am-5pm, T: 150-8'. Mark Leinberger, PH: 561-694-7982 or Jon Jacobus, PH: 954-772-1420.

Apr 17 1999 FL, Sarasota. 3rd Annual Doll & Bear Show. Municipal Auditorium, 801 N. Tamiami Trl., US Hwy. 41. SH: 10am-4pm, A: $4., 12 & under free with adult. Steve Schroeder, 3100 Harvest Ln., Kissimmee, FL 34744. PH: 407-957-6392 or FAX: 407-957-9427.

May 15-16 1999 FL, Orlando. Collectible Toy & Doll Super Show. Bahia Shrine Auditorium, 2300 Pembrook Dr. SH: Sat. 10am-5pm, Sun. 10am-4pm, T: 200-8'. Mark Leinberger, PH: 561-694-7982 or Jon Jacobus, PH: 954-772-1420.

Aug 14-15 1999 FL, Orlando. Collectible Toy & Doll Super Show. Bahia Shrine Auditorium, 2300 Pembrook Dr. SH: Sat. 10am-5pm, Sun. 10am-4pm, T: 200-8'. Mark Leinberger, PH: 561-694-7982 or Jon Jacobus, PH: 954-772-1420.

Aug 21 1999 FL, Ft. Lauderdale. South Florida Collectible Toy & Doll Super Show. Elks Exhibition Center, 700 NE 10th St., Pompano Beach. SH: 10am-5pm, T: 150-8'. Mark Leinberger, PH: 561-694-7982 or Jon Jacobus, PH: 954-772-1420.

Nov 6 1999 FL, Ft. Lauderdale. South Florida Collectible Toy & Doll Super Show. Elks Exhibition Center, 700 NE 10th St., Pompano Beach. SH: 10am-5pm, T: 150-8'. Mark Leinberger, PH: 561-694-7982 or Jon Jacobus, PH: 954-772-1420.

Nov 13-14 1999 FL, Orlando. Collectible Toy & Doll Super Show. Bahia Shrine Auditorium, 2300 Pembrook Dr. SH: Sat. 10am-5pm, Sun. 10am-4pm, T: 200-8'. Mark Leinberger, PH: 561-694-7982 or Jon Jacobus, PH: 954-772-1420.

GEORGIA

Jan 10 1999 GA, Marietta. Toy Show. Wyndham Garden Hotel, Atlanta North, I-75 North at Exit 112, Hwy. 120, South Loop. SH: 10am-5pm, T: 6'. KEBco Inc. Comic & Toy Emporium, PH: 770-471-6970.

Jan 16-17 1999 GA, Norcross. Greenberg'sGreat Train & Collectible Toy Show. North Atlanta Trade Ctr., Exit 38 off I-85. SH: Sat. 11am-5pm, Sun. 11am-4pm, A: $5., $2. ages 6-12, under 6 free. Greenberg Shows, Nan Turfle, 7566 Main St., Sykesville, MD 21784. PH: 410-795-7447.

Jan 16-17 1999 GA, Kennesaw. Collectible Show. Big Shanty Antique Mall, 1720 N. Roberts Rd., I-75, Exit 116, W. on Barrett Pky. SH: Sat. & Sun. 10am-5pm, T:

200-8', A: free. John Bennett, PH: 770-928-7068.

Feb 20-21 1999 GA, Kennesaw. Collectible Show. Big Shanty Antique Mall, 1720 N. Roberts Rd., I-75, Exit 116, W. on Barrett Pky. SH: Sat. & Sun. 10am-5pm, T: 200-8', A: free. John Bennett, PH: 770-928-7068.

Mar 7 1999 GA, Atlanta. The Joe & Marl Show. Airport Marriott, 4700 Best Rd. SH: 10am-3pm, A: $5., $2. under 12. Marl, PH: 941-751-6275 or Joe, PH: 213-953-6490.

Mar 20-21 1999 GA, Kennesaw. Collectible Show. Big Shanty Antique Mall, 1720 N. Roberts Rd., I-75, Exit 116, W. on Barrett Pky. SH: Sat. & Sun. 10am-5pm, T: 200-8', A: free. John Bennett, PH: 770-928-7068.

Apr 17-18 1999 GA, Kennesaw. Collectible Show. Big Shanty Antique Mall, 1720 N. Roberts Rd., I-75, Exit 116, W. on Barrett Pky. SH: Sat. & Sun. 10am-5pm, T: 200-8', A: free. John Bennett, PH: 770-928-7068.

May 15-16 1999 GA, Kennesaw. Collectible Show. Big Shanty Antique Mall, 1720 N. Roberts Rd., I-75, Exit 116, W. on Barrett Pky. SH: Sat. & Sun. 10am-5pm, T: 200-8', A: free. John Bennett, PH: 770-928-7068.

Jun 19-20 1999 GA, Kennesaw. Collectible Show. Big Shanty Antique Mall, 1720 N. Roberts Rd., I-75, Exit 116, W. on Barrett Pky. SH: Sat. & Sun. 10am-5pm, T: 200-8', A: free. John Bennett, PH: 770-928-7068.

Jul 17-18 1999 GA, Kennesaw. Collectible Show. Big Shanty Antique Mall, 1720 N. Roberts Rd., I-75, Exit 116, W. on Barrett Pky. SH: Sat. & Sun. 10am-5pm, T: 200-8', A: free. John Bennett, PH: 770-928-7068.

Aug 21-22 1999 GA, Kennesaw. Collectible Show. Big Shanty Antique Mall, 1720 N. Roberts Rd., I-75, Exit 116, W. on Barrett Pky. SH: Sat. & Sun. 10am-5pm, T: 200-8', A: free. John Bennett, PH: 770-928-7068.

Sep 18-19 1999 GA, Kennesaw. Collectible Show. Big Shanty Antique Mall, 1720 N. Roberts Rd., I-75, Exit 116, W. on Barrett Pky. SH: Sat. & Sun. 10am-5pm, T: 200-8', A: free. John Bennett, PH: 770-928-7068.

Oct 16-17 1999 GA, Kennesaw. Collectible Show. Big Shanty Antique Mall, 1720 N. Roberts Rd., I-75, Exit 116, W. on Barrett Pky. SH: Sat. & Sun. 10am-5pm, T: 200-8', A: free. John Bennett, PH: 770-928-7068.

Nov 20-21 1999 GA, Kennesaw. Collectible Show. Big Shanty Antique Mall, 1720 N. Roberts Rd., I-75, Exit 116, W. on Barrett Pky. SH: Sat. & Sun. 10am-5pm, T: 200-8', A: free. John Bennett, PH: 770-928-7068.

Dec 18-19 1999 GA, Kennesaw. Collectible Show. Big Shanty Antique Mall, 1720 N. Roberts Rd., I-75, Exit 116, W. on Barrett Pky. SH: Sat. & Sun. 10am-5pm, T: 200-8', A: free. John Bennett, PH: 770-928-7068.

HAWAII

Jul 23-25 1999 HI, Honolulu. Hawaii All-Collectors Show. Neal Blaisdell Hall, 777 Ward Ave. SH: Fri. 4pm-9pm, Sat. 11am-9pm, Sun. 11am-5pm, T: 200. Wayne Maeda, PO Box 61704, Honolulu, HI 96839. PH: 808-941-9754.

ILLINOIS

Jan 2-3 1999 IL, Joliet. 2nd Annual Collectible Show. Harwood Post, I-80 & Larkin Ave. S. SH: Sat. 10am-3pm,

Sun. 8am-1pm, T: 60. Edward Streich, 24728 W. Hemphill Dr., Elwood, IL 60421. PH: 815-478-4362.

Jan 2 1999 IL, Bloomington. Toys-N-Collectibles Show. Ramada Inn, N. Veterans Pky. SH: 10am-4pm, T: 40. The Front Porch, Cindy, 515 Mill St. Ste. 107, Mt. Zion, IL 62544. PH: 217-864-2707.

Jan 3 1999 IL, Grayslake. Skip's Car & Truck Parts Swap Meet & Die Cast Show. Lake Cty. Fairgrounds, Rt. 45 & Rt. 120. SH: 8am-4pm, A: $5. Skip's Shows, PO Box 88266, Carol Stream, IL 60188. PH: 630-682-8792 or 800-050-7369.

Jan 3 1999 IL, Springfield. Toys-N-Collectibles Show. Ramada Inn South, I-55 to 6th St. Exit. SH: 10am-4pm, T: 40. The Front Porch, Cindy, 515 Mill St. Ste. 107, Mt. Zion, IL 62544. PH: 217-864-2707.

Jan 3 1999 IL, Oak Lawn. Toy, Beanies, Hot Wheels & Sports Card Show. Fatima K of C, 5830 W. 95th St. SH: 9am-3pm, T: 40. John Leary, 9522 W. Shore Dr., Oak Lawn, IL 60453. PH: 708-423-1758.

Jan 9 1999 IL, Orland Park. Toys, Beanies, Hot Wheels & Sports Cards Show. Civic Ctr., 1 blk. W. of LaGrange Rd. (Rt. 45) at 147th St. SH: 9am-3pm, T: 85. John Leary, 9522 W. Shore Dr., Oak Lawn, IL 60453. PH: 708-423-1758.

Jan 10 1999 IL, Bridgeview. If Its Got Wheels. John A Oremus Community Ctr., 7900 S. Oketo Ave. SH: 8am-2pm, T: 75. James LaCoco, 231 Englewood Ave., Bellwood, IL 60104. PH: 708-544-1975.

Jan 24 1999 IL, Orland Park. Toys, Beanies, Hot Wheels & Sports Cards Show. Civic Ctr., 1 blk. W. of LaGrange Rd. (Rt. 45) at 147th St. SH: 9am-3pm, T: 85. John Leary, 9522 W. Shore Dr., Oak Lawn, IL 60453. PH: 708-423-1758.

Jan 30-31 1999 IL, Collinsville. Greenberg's Great Train & Collectible Toy Show. Gateway Ctr., Eastport Plaza Dr. SH: 11am-5pm, Sun. 11am-4pm, A: $5., $2. ages 6-12, under 6 free. Greenberg Shows, Nan Turfle, 7566 Main St., Sykesville, MD 21784. PH: 410-795-7447.

Feb 7 1999 IL, Westmont. Antique, Collectible, Modern Doll, Teddy Bear & Miniature Show. Inland Meeting & Expo Ctr., 400 E. Ogden. SH: 9am-4pm, T: 100. Gigi's Dolls & Sherry's Teddy Bears, Inc., 6029 N. Northwest Hwy., Chicago, IL 60631. PH: 773-594-1540 or FAX: 773-594-1710.

Feb 7 1999 IL, Worth. 10th Annual Automotive Literature & Automobilia Swap Meet. Park District, 3 blks. W. on 115th at Harlem Ave. SH: 8am-2pm, T: 100. Chicago On Wheels Motorsports, Thomas Lima, PO Box 586, Cedar Lake, IN 46303. PH: 219-374-8780 or FAX: 219-374-5962.

Feb 7 1999 IL, Oak Lawn. Toy, Beanies, Hot Wheels & Sports Card Show. Fatima K of C, 5830 W. 95th St. SH: 9am-3pm, T: 40. John Leary, 9522 W. Shore Dr., Oak Lawn, IL 60453. PH: 708-423-1758.

Feb 13 1999 IL, Orland Park. Toys, Beanies, Hot Wheels & Sports Cards Show. Civic Ctr., 1 blk. W. of LaGrange Rd. (Rt. 45) at 147th St. SH: 9am-3pm, T: 85. John Leary, 9522 W. Shore Dr., Oak Lawn, IL 60453. PH: 708-423-1758.

Feb 28 1999 IL, Milan. Quad-City Antique Toy & Doll Show.

Community Center, Hwy. 67. SH: 9am-3pm, T: 125. Ron Aust, 3715 Volquardsen, Davenport, IA 52806. PH: 319-391-3579.

Feb 28 1999 IL, Orland Park. Toys, Beanies, Hot Wheels & Sports Cards Show. Civic Ctr., 1 blk. W. of LaGrange Rd. (Rt. 45) at 147th St. SH: 9am-3pm, T: 85. John Leary, 9522 W. Shore Dr., Oak Lawn, IL 60453. PH: 708-423-1758.

Mar 7 1999 IL, Hillside. Super Model Car Sunday Winter Swap Meet. Holiday Inn, I-290 at Wolf Rd. SH: 9am-3pm, T: 175. Chicago On Wheels Motorsports, Thomas Lima, PO Box 586, Cedar Lake, IN 46303. PH: 219-374-8780 or FAX: 219-374-5962.

Mar 7 1999 IL, Oak Lawn. Toy, Beanies, Hot Wheels & Sports Card Show. Fatima K of C, 5830 W. 95th St. SH: 9am-3pm, T: 40. John Leary, 9522 W. Shore Dr., Oak Lawn, IL 60453. PH: 708-423-1758.

Mar 13 1999 IL, Orland Park. Toys, Beanies, Hot Wheels & Sports Cards Show. Civic Ctr., 1 blk. W. of LaGrange Rd. (Rt. 45) at 147th St. SH: 9am-3pm, T: 85. John Leary, 9522 W. Shore Dr., Oak Lawn, IL 60453. PH: 708-423-1758.

Mar 14 1999 IL, Chicago. Antique Collectible Toy Show. Brother Rice High School, 10001 S. Pulaski. SH: 9am-4pm. Tom Hornik, PH: 708-923-6358.

Mar 14 1999 IL, Rolling Meadows. Beanie Expo Spring Fling. Holiday Inn, on Algonquin Rd. (Rt. 62) & of 53. SH: 10am-4pm, A: $5. Carol Ann, PO Box 6266, Buffalo Grove, IL 60089. PH: 847-520-1412.

Mar 28 1999 IL, Orland Park. Toys, Beanies, Hot Wheels & Sports Cards Show. Civic Ctr., 1 blk. W. of LaGrange Rd. (Rt. 45) at 147th St. SH: 9am-3pm, T: 85. John Leary, 9522 W. Shore Dr., Oak Lawn, IL 60453. PH: 708-423-1758.

Apr 4 1999 IL, Oak Lawn. Toy, Beanies, Hot Wheels & Sports Card Show. Fatima K of C, 5830 W. 95th St. SH: 9am-3pm, T: 40. John Leary, 9522 W. Shore Dr., Oak Lawn, IL 60453. PH: 708-423-1758.

Apr 10 1999 IL, Orland Park. Toys, Beanies, Hot Wheels & Sports Cards Show. Civic Ctr., 1 blk. W. of LaGrange Rd. (Rt. 45) at 147th St. SH: 9am-3pm, T: 85. John Leary, 9522 W. Shore Dr., Oak Lawn, IL 60453. PH: 708-423-1758.

Apr 25 1999 IL, Saint Charles. Antique Toy & Doll World Shows. Kane Cty. Fairgrounds, Rt. 64 & Randall Rd. SH: 8am-4pm. Antique World Shows, PO Box 34509, Chicago, IL 60634. PH: 847-526-1645 or FAX: 847-526-7416.

Apr 25 1999 IL, Orland Park. Toys, Beanies, Hot Wheels & Sports Cards Show. Civic Ctr., 1 blk. W. of LaGrange Rd. (Rt. 45) at 147th St. SH: 9am-3pm, T: 85. John Leary, 9522 W. Shore Dr., Oak Lawn, IL 60453. PH: 708-423-1758.

May 2 1999 IL, Oak Lawn. Toy, Beanies, Hot Wheels & Sports Card Show. Fatima K of C, 5830 W. 95th St. SH: 9am-3pm, T: 40. John Leary, 9522 W. Shore Dr., Oak Lawn, IL 60453. PH: 708-423-1758.

May 8 1999 IL, Orland Park. Toys, Beanies, Hot Wheels & Sports Cards Show. Civic Ctr., 1 blk. W. of LaGrange Rd. (Rt. 45) at 147th St. SH: 9am-3pm, T: 85. John Leary, 9522 W. Shore Dr., Oak Lawn, IL 60453. PH: 708-423-1758.

May 23 1999 IL, Orland Park. Toys, Beanies, Hot Wheels & Sports Cards Show. Civic Ctr., 1 blk. W. of LaGrange Rd. (Rt. 45) at 147th St. SH: 9am-3pm, T: 85. John Leary, 9522 W. Shore Dr., Oak Lawn, IL 60453. PH: 708-423-1758.

Jun 6 1999 IL, Oak Lawn. Toy, Beanies, Hot Wheels & Sports Card Show. Fatima K of C, 5830 W. 95th St. SH: 9am-3pm, T: 40. John Leary, 9522 W. Shore Dr., Oak Lawn, IL 60453. PH: 708-423-1758.

Jun 12 1999 IL, Orland Park. Toys, Beanies, Hot Wheels & Sports Cards Show. Civic Ctr., 1 blk. W. of LaGrange Rd. (Rt. 45) at 147th St. SH: 9am-3pm, T: 85. John Leary, 9522 W. Shore Dr., Oak Lawn, IL 60453. PH: 708-423-1758.

Jun 27 1999 IL, Saint Charles. Antique Toy & Doll World Shows. Kane Cty. Fairgrounds, Rt. 64 & Randall Rd. SH: 8am-4pm.

Antique World Shows, PO Box 34509, Chicago, IL 60634. PH: 847-526-1645 or FAX: 847-526-7416.

Jun 27 1999 IL, Orland Park. Toys, Beanies, Hot Wheels & Sports Cards Show. Civic Ctr., 1 blk. W. of LaGrange Rd. (Rt. 45) at 147th St. SH: 9am-3pm, T: 85. John Leary, 9522 W. Shore Dr., Oak Lawn, IL 60453. PH: 708-423-1758.

Jul 25 1999 IL, Worth. Super Model Car Sunday Annual Summer Swap Meet. Park District, 3 blks. W. on 115th at Harlem Ave. SH: 8am-2pm, T: 100. Chicago On Wheels Motorsports, Thomas Lima, PO Box 586, Cedar Lake, IN 46303. PH: 219-374-8780 or FAX: 219-374-5962.

Sep 26 1999 IL, Westmont. Antique, Collectible, Modern Doll, Teddy Bear & Miniature Show. Inland Meeting & Expo Ctr., 400 E. Ogden. SH: 9am-4pm, T: 100. Gigi's Dolls & Sherry's Teddy Bears, Inc., 6029 N. Northwest Hwy., Chicago, IL 60631. PH: 773-594-1540 or FAX: 773-594-1710.

Oct 17 1999 IL, Hillside. Super Model Car Sunday Fall Swap Meet. Holiday Inn, I-290 at Wolf Rd. SH: 9am-3pm, T: 175. Chicago On Wheels Motorsports, Thomas Lima, PO Box 586, Cedar Lake, IN 46303. PH: 219-374-8780 or FAX: 219-374-5962.

Oct 24 1999 IL, Saint Charles. Antique Toy & Doll World Shows. Kane Cty. Fairgrounds, Rt. 64 & Randall Rd. SH: 8am-4pm. Antique World Shows, PO Box 34509, Chicago, IL 60634. PH: 847-526-1645 or FAX: 847-526-7416.

Oct 24 1999 IL, Worth. Automotive Toy & Model Car Annual Swap Meet. Park District, 3 blks. W. on 115th at Harlem Ave. SH: 8am-2pm, T: 100. Chicago On Wheels Motorsports, Thomas Lima, PO Box 586, Cedar Lake, IN 46303. PH: 219-374-8780 or FAX: 219-374-5962.

Dec 11 1999 IL, Lansing. Hot Wheels, Model Car & Toy Car Swap Meet. Trinity Lutheran School Gym, 18144 Glen Terrace. SH: 8am-2pm, T: 75. Chicago On Wheels Motorsports, Thomas Lima, PO Box 586, Cedar Lake, IN 46303. PH: 219-374-8780 or FAX: 219-374-5962.

INDIANA

Mar 13 1999 IN, Ft. Wayne. Train & Collectable Toy Show. The Lantern, 4420 Ardmore. SH: 11am-4pm, T: 160. Sally Valiton, 7112 Baer Rd., Ft. Wayne, IN 46809. PH: 219-747-4485.

May 2 1999 IN, Crown Point. Hot Wheels & Model Car Annual Swap Meet. Lake Country Indiana Fairgrounds, 4-H Bldg., 833 S. Court St. SH: 8am-2pm, T: 75. Chicago On Wheels Motorsports, Thomas Lima, PO Box 586, Cedar Lake, IN 46303. PH: 219-374-8780 or FAX: 219-374-5962.

May 16 1999 IN, La Porte. Just Toys-Toy Show. Co. Fairgrounds, Hwy. 2 West. SH: 9am-3pm, T: 275, A: $3. Jeff Plante, PH: 219-362-6958.

Jun 20 1999 IN, Highlands. Hot Wheels & Model Car & Toy Car Swap Meet. Blue-Top Drive-In, 8801 Indianapolis Blvd., Rt. 41. SH: 8am-2pm, T: 75. Chicago On Wheels Motorsports, Thomas Lima, PO Box 586, Cedar Lake, IN 46303. PH: 219-374-8780 or FAX: 219-374-5962.

Nov 21 1999 IN, Crown Point. Hot Wheels & Model Car Annual Swap Meet. Lake Country Indiana Fairgrounds, 4-H Bldg., 833 S. Court St. SH: 8am-2pm, T: 75. Chicago On Wheels Motorsports, Thomas Lima, PO Box 586, Cedar Lake, IN 46303. PH: 219-374-8780 or FAX: 219-374-5962.

IOWA

Mar 14 1999 IA, Maquoketa. 17th Annual Doll, Toy & Bear Show. Jackson Cty. Fairgrounds, Jct. Hwys. 62 & 64. SH: 9am-4pm. Dora Pitts, 4697 155th St., Clinton, IA 52732. PH: 319-242-0139.

May 8 1999 IA, Des Moines. 11th Annual Doll, Toy & Bear Show. Soccer & Sports Ctr., 5406 Merle Hay Rd., off I-80, Exit 131 N. SH: 9am-4pm. Dora Pitts, 4697 155th St., Clinton, IA 52732. PH: 319-242-0139.

Sep 12 1999 IA, Maquoketa. Annual Fall Doll, Toy & Bear Show. Jackson Cty. Fairgrounds, Jct. Hwys. 62 & 64. SH: 9am-4pm. Dora Pitts, 4697 155th St., Clinton, IA 52732. PH: 319-242-0139.

Oct 9 1999 IA, Des Moines. Annual Fall Doll, Toy & Bear Show. Soccer & Sports Ctr., 5406 Merle Hay Rd., off I-80, Exit 131 N. SH: 9am-4pm. Dora Pitts, 4697 155th St., Clinton, IA 52732. PH: 319-242-0139.

Nov 14 1999 IA, Davenport. 15th Annual Wonderland Doll, Toy & Bear Show. Mississippi Valley Fairgrounds, 2815 W. Locust St. SH: 9am-4pm. Dora Pitts, 4697 155th St., Clinton, IA 52732. PH: 319-242-0139.

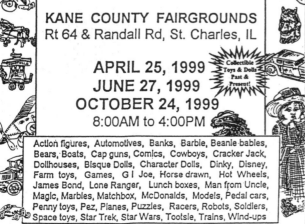

KANSAS

Mar 14 1999 KS, Kansas City. Annual Toy & Doll Spring Show. Jack Reardon Civic Ctr., 5th & Minnesota. SH: 10am-4pm, A: $3. Judy Condray, 1005 W. 11th, Concordia, KS 66901. PH: 785-243-3774.

Sep 5 1999 KS, Kansas City. Annual Toy & Doll Fall Show. Jack Reardon Civic Ctr., 5th & Minnesota. SH: 10am-4pm, A: $3. Bob Condray, PH: 785-455-3440 or Scott & Mindy Condray, PH: 785-243-7724.

MARYLAND

Jan 23-24 1999 MD, Upper Marlboro. Greenberg's Great Train & Collectible Toy Show. Show Place Arena, Rts. 4 & 301. SH: Sat. 11am-4pm, Sun. 11am-4pm, A: $5., $2. ages 6-12, under 6 free. Greenberg Shows, Nan Turfle, 7566 Main St., Sykesville, MD 21784. PH: 410-795-7447.

Jan 30 1999 MD, Baltimore. Greater Antique & Collectible Toy Show. Tall Cedars Hall, 2501 Putty Hill Ave. SH: 9am-3pm, T: 140. Raymond Bosley & David Hart, 3010 Merrymans Mill Rd., Phoenix, MD 21131. PH: 410-628-7979 or 433-4278.

Mar 20 1999 MD, Baltimore. Greater Antique & Collectible Toy Show. Tall Cedars Hall, 2501 Putty Hill Ave. SH: 9am-3pm, T: 140. Raymond Bosley & David Hart, 3010 Merrymans Mill Rd., Phoenix, MD 21131. PH: 410-628-7979 or 433-4278.

Mar 20-21 1999 MD, Timonium. Greenberg's Great Train & Collectible Toy Show. State Fairgrounds, I-83, Exit 16A or 17. SH: Sat. 11am-5pm, Sun. 11am-4pm, A: $5., $2. ages 6-12, under 6 free. Greenberg Shows, Nan Turfle, 7566 Main St., Sykesville, MD 21784. PH: 410-795-7447.

Mar 28 1999 MD, Chevy Chase. Collectible Toy & Sports Card Show. Leland Community Ctr., 4301 Leland St. SH: 10am-3pm, T: 35-6', A: free. Claron McDaniel, 4301 Leland, Chevy Chase, MD 20815. PH: 301-652-2249.

May 16 1999 MD, Freeland. Diecast & Collectible Vehicle Toy Show. Morris Meadows Historic Preservation Museum, 1523 Freeland Rd. SH: 9am-2pm. Joe Golabiewski, 12317 Harford Rd., Kingsville, MD 21087. PH: 410-592-5854 after 6pm.

MASSACHUSETTS

Feb 14 1999 MA, Dedham. Annual Cabin Fever Show. Holiday Inn, Rt. 1, Exit 15A. SH: 9:30am-3pm. Mrs. Devlin, PH: 508-379-9733.

Feb 14 1999 MA, South Attleboro. Toys, Comic Book, Non-Sport Cards & Coll. Show. K of C, 304 Highland Ave. SH: 10am-3pm. Harry Martin, 936 South St., Wrentham, MA 02093. PH: 508-384-8491.

Mar 14 1999 MA, Auburn. All Barbie Show. Ramada Inn, 624 Southbridge St. Exit 10 off Mass Pike. SH: 10am-3pm. PH: 978-342-8292 or 342-2265.

Mar 21 1999 MA, Dedham. Annual Spring Show. Holiday Inn, Rt. 1, Exit 15A. SH: 9:30am-3pm. Mrs. Devlin, PH: 508-379-9733.

Mar 28 1999 MA, South Attleboro. Toys, Comic Book, Non-Sport Cards & Coll. Show. K of C, 304 Highland Ave. SH: 10am-3pm. Harry Martin, 936 South St., Wrentham, MA 02093. PH: 508-384-8491.

Apr 11 1999 MA, Peabody. 12th East Mass Model & Toys Show. Holiday Inn, 1 Newbury St. (Rt. 1 N.). SH: 9am-2pm, T: 60. Michael Zelikson, PH: 781-631-9677.

Apr 18 1999 MA, Dedham. Northeast Toy Soldier Show. Holiday Inn, Exit 15A, I-95 (Rt. 128), Jct. Rt. 1 S. SH: 9am-3pm, T: 100. Arley Pett, 12 Beach Rd., Gloucester, MA 01930. PH: 978-283-2612.

May 11-16 1999 MA, Brimfield. Antique & Collectibles Show. Central Park, Rt. 20. SH: 6am-6pm. Patricia Waite, PO Box 224, Brimfield, MA 01095. PH: 413-596-9257.

May 16 1999 MA, South Attleboro. Toys, Comic Book, Non-Sport Cards & Coll. Show. K of C, 304 Highland Ave. SH: 10am-3pm. Harry Martin, 936 South St., Wrentham, MA 02093. PH: 508-384-8491.

Jun 27 1999 MA, South Attleboro. Toys, Comic Book, Non-Sport Cards & Coll. Show. K of C, 304 Highland Ave. SH: 10am-3pm. Harry Martin, 936 South St., Wrentham, MA 02093. PH: 508-384-8491.

Jul 6-11 1999 MA, Brimfield. Antique & Collectibles Show. Central Park, Rt. 20. SH: 6am-6pm. Patricia Waite, PO Box 224, Brimfield, MA 01095. PH: 413-596-9257.

Aug 15 1999 MA, South Attleboro. Toys, Comic Book, Non-Sport Cards & Coll. Show. K of C, 304 Highland Ave. SH: 10am-3pm. Harry Martin, 936 South St., Wrentham, MA 02093. PH: 508-384-8491.

Sep 14-19 1999 MA, Brimfield. Antique & Collectibles Show. Central Park, Rt. 20. SH: 6am-6pm. Patricia Waite, PO Box 224, Brimfield, MA 01095. PH: 413-596-9257.

Sep 26 1999 MA, South Attleboro. Toys, Comic Book, Non-Sport Cards & Coll. Show. K of C, 304 Highland Ave. SH: 10am-3pm. Harry Martin, 936 South St., Wrentham, MA 02093. PH: 508-384-8491.

Oct 17 1999 MA, Dedham. Northeast Toy Soldier Show. Holiday Inn, Exit 15A, I-95, (Rt. 128), Jct. Rt. 1 S. SH: 9am-3pm, T: 100. Arley Pett, 12 Beach Rd., Gloucester, MA 01930. PH: 978-283-2612.

Oct 24 1999 MA, Peabody. 13th East Mass Model & Toys Show. Holiday Inn, 1 Newbury St. (Rt. 1 N.). SH: 9am-2pm, T: 60. Michael Zelikson, PH: 781-631-9677.

Nov 14 1999 MA, South Attleboro. Toys, Comic Book, Non-Sport Cards & Coll. Show. K of C, 304 Highland Ave. SH: 10am-3pm. Harry Martin, 936 South St., Wrentham, MA 02093. PH: 508-384-8491.

Nov 21 1999 MA, Sturbridge. New England Ornament & Collectibles Secondary Market Expo. Host Hotel, Rt. 20. SH: 10am-3pm, T: 100. Patricia Waite, PO Box 515, Wilbraham, MA 01095. PH: 413-596-9257.

Dec 19 1999 MA, South Attleboro. Toys, Comic Book, Non-Sport Cards & Coll. Show. K of C, 304 Highland Ave. SH: 10am-3pm. Harry Martin, 936 South St., Wrentham, MA 02093. PH: 508-384-8491.

MICHIGAN

Jan 3 1999 MI, Livonia. Toy-Fest. Knights of Columbus Hall, 19801 Farmington Rd. (btw. 7 & 8 Mile Rd.). SH: 10am-3pm, A: $3. Lauren Becker, 3411 Heron Pt. Ct., Waterford, MI 48328. PH: 248-706-0886 or FAX: 248-738-9032.

Jan 10 1999 MI, Plymouth. Beanie Baby Show. Cultural Center, 525 Farmer St. SH: 11am-3pm, A: $5., $2. ages 4-12. R.R. Promotions, Inc., PO Box 6094, Plymouth, MI 48170. PH/FAX: 734-455-2110.

Jan 16 1999 MI, Plymouth. Train Show. Cultural Ctr., 525 Farmer St. SH: 11am-5pm. R.R. Promos., Inc., PO Box 6094, Plymouth, MI 48170. PH/FAX: 734-455-2110.

Jan 17 1999 MI, Plymouth. Toy Show. Cultural Ctr., 525 Farmer St. SH: 11am-5pm. R.R. Promos., Inc., PO Box 6094, Plymouth, MI 48170. PH/FAX: 734-455-2110.

Jan 24 1999 MI, Warren. Toy Show. Macomb Community College Sport & Expo Center (Field House), 12 Mile at Hayes Rd. SH: 9am-2:30pm, T: 300. Sal Carrato, c/o D.A.A.M., PO Box 3488, Centerline, MI 48015. PH: 810-247-7464.

Feb 14 1999 MI, Plymouth. We Love Barbie Doll Show. Cultural Center, 525 Farmer St. SH: 11am-4pm, A: $5., $2. ages 4-12. R.R. Promotions, Inc., PO Box 6094, Plymouth, MI 48170. PH/FAX: 734-455-2110.

Mar 14 1999 MI, Plymouth. Beanie Baby Show. Cultural Center, 525 Farmer St. SH: 11am-3pm, A: $5., $2. ages 4-12. R.R. Promotions, Inc., PO Box 6094, Plymouth, MI 48170. PH/FAX: 734-455-2110.

Mar 21 1999 MI, Livonia. Toy-Fest. Knights of Columbus Hall, 19801 Farmington Rd. (btw. 7 & 8 Mile Rd.). SH: 10am-3pm, A: $3. Lauren Becker, 3411 Heron Pt. Ct., Waterford, MI 48328. PH: 248-706-0886 or FAX: 248-738-9032.

Apr 11 1999 MI, Plymouth. Collectible Toy & Model Kit Show. Cultural Center, 525 Farmer St. SH: 11am-4pm, A: $5., $2. ages 4-12. R.R. Promotions, Inc., PO Box 6094, Plymouth, MI 48170. PH/FAX: 734-455-2110.

May 2 1999 MI, Plymouth. Beanie Baby Show. Cultural Center, 525 Farmer St. SH: 11am-3pm, A: $5. ages 4-12. R.R. Promotions, Inc., PO Box 6094, Plymouth, MI 48170. PH/FAX: 734-455-2110.

Jun 5-6 1999 MI, Midland. Antique & Collectible Festival. Fairgrounds, US 10 at Eastman Ave. SH: Sat. 7am-9pm, Sun. 7am-4pm, T: 800. Hugh Smith, 2156 Rudy Ct., Midland, MI 48642. PH: 517-687-9001.

Jul 11 1999 MI, Plymouth. Beanie Baby Show. Cultural Center, 525 Farmer St. SH: 11am-3pm, A: $5., $2. ages 4-12. R.R. Promotions, Inc., PO Box 6094, Plymouth, MI 48170. PH/FAX: 734-455-2110.

Jul 24-25 1999 MI, Midland. Antique & Collectible Festival. Fairgrounds, US 10 at Eastman Ave. SH: Sat. 7am-9pm, Sun. 7am-4pm, T: 800. Hugh Smith, 2156 Rudy Ct., Midland, MI 48642. PH: 517-687-9001.

Aug 8 1999 MI, Plymouth. Beanie Baby Show. Cultural Center, 525 Farmer St. SH: 11am-3pm, A: $5., $2. ages 4-12. R.R. Promotions, Inc., PO Box 6094, Plymouth, MI 48170. PH/FAX: 734-455-2110.

Sep 12 1999 MI, Plymouth. Beanie Baby Show. Cultural Center, 525 Farmer St. SH: 11am-5pm, A: $5., $2. ages 4-12. R.R. Promotions, Inc., PO Box 6094, Plymouth, MI 48170. PH/FAX: 734-455-2110.

Sep 19 1999 MI, Plymouth. Model Railroad & Toy Train

Show. Cultural Center, 525 Farmer St. SH: 11am-4pm, A: $5., $2. ages 4-12. R.R. Promotions, Inc., PO Box 6094, Plymouth, MI 48170. PH/FAX: 734-455-2110.

Sep 25-26 1999 MI, Midland. Antique & Collectible Festival. Fairgrounds, US 10 at Eastman Ave. SH: Sat. 7am-9pm, Sun. 7am-4pm, T: 800. Hugh Smith, 2156 Rudy Ct., Midland, MI 48642. PH: 517-687-9001.

Oct 3 1999 MI, Plymouth. Collectible Toy & Model Kit Show. Cultural Center, 525 Farmer St. SH: 11am-4pm, A: $5., $2. ages 4-12. R.R. Promotions, Inc., PO Box 6094, Plymouth, MI 48170. PH/FAX: 734-455-2110.

Oct 10 1999 MI, Plymouth. Beanie Baby Show. Cultural Center, 525 Farmer St. SH: 11am-3pm, A: $5., $2. ages 4-12. R.R. Promotions, Inc., PO Box 6094, Plymouth, MI 48170. PH/FAX: 734-455-2110.

Nov 7 1999 MI, Plymouth. Beanie Baby Show. Cultural Center, 525 Farmer St. SH: 11am-3pm, A: $5., $2. ages 4-12. R.R. Promotions, Inc., PO Box 6094, Plymouth, MI 48170. PH/FAX: 734-455-2110.

Dec 12 1999 MI, Plymouth. Antique & Collectible Doll Show. Cultural Center, 525 Farmer St. SH: 11am-4pm, A: $5., $2. ages 4-12. R.R. Promotions, Inc., PO Box 6094, Plymouth, MI 48170. PH/FAX: 734-455-2110.

MINNESOTA

Jan 17 1999 MN, Brooklyn Center. 18th Annual Winter Doll, Bear & Miniature Show. Earle Brown Heritage Ctr., 6155 Earle Brown Dr. (I-694 & Hwy. 100). SH: 10am-4pm. Carol's Doll House, 10761 University Ave. NE, Blaine, MN 55434. PH: 612-755-7475.

Feb 14 1999 MN, Bloomington. 28th Bi-Annual Teddy, Doll & Toy Show. Days Inn Motel, 8401 Cedar Ave. S. (77). SH: 10am-4pm. PH: 612-898-3526.

Feb 27 1999 MN, Le Sueur. Toy Show. Park Elementary School, 105 N. 5th St. SH: 9am-4pm. Jay Wieland, 213 Bridge St., Le Sueur, MN 56058. PH: 507-665-6312.

MISSOURI

May 9 1999 MO, St. Louis. The Joe & Marl Show. Renaissance-St. Louis Airport, 9801 Natural Bridge Rd. SH: 10am-3pm, A: $5., $2. under 12. Marl, PH: 941-751-6275 or Joe, PH: 213-953-6490.

NEVADA

Apr 11 1999 NV, Las Vegas. The Joe & Marl Show. The Gold Coast, 4000 W. Flamingo Rd. SH: 10am-3pm, A: $5., $2. under 12. Marl, PH: 941-751-6275 or Joe, PH: 213-953-6490.

NEW JERSEY

Jan 1-2 1999 NJ, E. Brunswick. New Year's Day Toy, Sports Card & Collectibles Show. Ramada Inn, Rt. 18 S. at NJ Tpke. Exit 9. SH: Fri. 11am-5pm, Sat. 10am-4pm, A: $2., under 6 free. Sallie Natowitz, PO Box 796, Matawan, NJ 07747. PH: 732-583-7915.

Jan 3 1999 NJ, East Hanover. Greater Northeast Big Toy Event. Ramada Hotel, 130 Rt. 10 W. SH: 10am-4pm, T: 150. Yolanda Stanczyk, PO Box 368, Tannersville, PA 18372. PH: 717-620-2422.

Jan 9 1999 NJ, North Branch. Toys, Sports Cards & Collectibles Show. Fire House, Rt. 28, next to Raritan Valley Comunity College. SH: 10am-4pm, A: $1., under 6 free. Sallie Natowitz, PO Box 796, Matawan, NJ 07747. PH: 732-583-7915.

Jan 10 1999 NJ, Bellmawr. Automotive Toys & Models NASCAR Collectibles Show. Howard Johnson Motor-lodge, Rt. 168 (Black Horse Pike) at NJ Tpke. Exit 3. SH: 9am-3pm, T: 50, A: $2., under 12 free. High Speed Promotions, Steven Rosenzweig, 349 Windsor Dr., Cherry Hill, NJ 08002. PH: 609-667-6808.

Jan 16 1999 NJ, Colts Neck. Doll & Bear Show. Firehouse #2, Conover Rd. just off Rt. 34. SH: 10am-4pm, A: $1.50, under 6 free. Sallie Natowitz, PO Box 796, Matawan, NJ 07747. PH: 732-583-7915.

Jan 16 1999 NJ, Hazlet. Beanie Baby Super Show. Hazlet Hotel, 2870 Hwy. 35 (Pky. Exit 117 to Rt. 35 S. 2 mi.). SH: 9:30am-3:30pm, T: 45. Collectors Showcase, Sandy Grecco, PO Box 189, Atlantic Highlands, NJ 07716. PH: 732-291-1632.

Jan 17 1999 NJ, East Hanover. Greater Northeast Doll & Teddy Bear Show. Ramada Hotel, 130 Rt. 10 W. SH: 10am-4pm, T: 150. Yolanda Stanczyk, PO Box 368, Tannersville, PA 18372. PH: 717-620-2422.

Jan 23 1999 NJ, Kenilworth. Beanie Baby Super Show. Kenilworth Inn, off GSP Exit 138, turn L. SH: 9:30am-3:30pm, T: 45. Collectors Showcase, Sandy Grecco, PO Box 189, Atlantic Highlands, NJ 07716. PH: 732-291-1632.

Jan 24 1999 NJ, Wayne. The Great Bergen Passaic Toy & Train Show. P.A.L. Hall. SH: 9am-2pm. Vic, PH: 516-653-8133.

Jan 24 1999 NJ, Princeton. Doll & Bear Show. Holiday Inn, 4355 Rt. 1 S. at Ridge Rd. SH: 10am-4pm, A: $2., under 6 free. Sallie Natowitz, PO Box 796, Matawan, NJ 07747. PH: 732-583-7915.

Jan 24 1999 NJ, Toms River. Beanie Baby Super Show. Holiday Inn, 290 Hwy. 34 E. SH: 9:30am-3:30pm, T: 35. Collectors Showcase, Sandy Grecco, PO Box 189, Atlantic Highlands, NJ 07716. PH: 732-291-1632.

Feb 7 1999 NJ, Livingston. NJ Doll Show. Travelodge, 550 W. Mt. Pleasant Ave. SH: 10am-4pm. Key Promotions Ltd., PH: 908-233-7949.

Feb 7 1999 NJ, Somerset. Doll & Bear Show. Ramada Inn, Weston Canal Rd. at I-287, Exit 12. SH: 10am-4pm, A: $2., under 6 free. Sallie Natowitz, PO Box 796, Matawan, NJ 07747. PH: 732-583-7915.

Feb 7 1999 NJ, Hazlet. Beanie Baby Super Show. Hazlet Hotel, 2870 Hwy. 35 (Pky. Exit 117 to Rt. 35 S. 2 mi.). SH: 9:30am-3:30pm, T: 45. Collectors Showcase, Sandy Grecco, PO Box 189, Atlantic Highlands, NJ 07716. PH: 732-291-1632.

Feb 13 1999 NJ, North Branch. Toys, Sports Cards & Collectibles Show. Fire House, Rt. 28, next to Raritan Valley Comunity College. SH: 10am-4pm, A: $1., under 6 free. Sallie Natowitz, PO Box 796, Matawan, NJ 07747. PH: 732-583-7915.

Feb 14 1999 NJ, Hazlet. Collectible Toys, Comic Books, Lineups & Action Figure Show. Hazlet Hotel, 2870 Hwy. 35 (Pky. Exit 117 to Hwy. 35 S. 2 mi.). SH: 9:30am-3:30pm, T: 50. Collectors Showcase, Sandy Grecco, PO Box 189, Atlantic Highlands, NJ 07716. PH: 732-291-1632.

Feb 20-21 1999 NJ, Cherry Hill. TV, Film, Toys & Autograph Collectibles Show. Holiday Inn, Rt. 70 E. SH: 10am-4pm, A: $6. JD Prods., PO Box 726, Cherry Hill, NJ 08003. PH: 609-795-0436 or FAX: 609-795-1475.

Feb 20-21 1999 NJ, Ocean Township. Toy, Sports Card, Non-Sports Card, Comic Book & Collectible Show. Seaview Square Mall, Rts. 35 & 66, Pky. Exit 102 S. or 100A N. to Rt. 66 E. for 2.5 mi. SH: Sat. 10am-9pm, Sun. 11am-5pm, T: 70-8', A: free. Collectors Showcase, Sandy Grecco, PO Box 189, Atlantic Highlands, NJ 07716. PH: 732-291-1632.

Feb 27-28 1999 NJ, Edison. Greenberg's Great Train & Collectible Toy Show. NJ Convention & Expo Ctr., 97 Sunfield Ave. SH: Sat. 11am-5pm, Sun. 11am-4pm, A: $5., $2. ages 6-12, under 6 free. Greenberg Shows, Nan Turfle, 7566 Main St., Sykesville, MD 21784. PH: 410-795-7447.

Feb 27 1999 NJ, Kenilworth. Beanie Baby Super Show. Kenilworth Inn, off GSP Exit 138, turn L. SH: 9:30am-3:30pm, T: 45. Collectors Showcase, Sandy Grecco, PO Box 189, Atlantic Highlands, NJ 07716. PH: 732-291-1632.

Feb 28 1999 NJ, Livingston. Toy Show. Travelodge, 550 W. Mt. Pleasant Ave. (Rt. 10). SH: 10am-4pm, A: $4.50, $1. under 12. Key Promotions Ltd., PO Box 51, Metuchen, NJ 08840. PH: 908-756-2385 or 233-7949.

Feb 28 1999 NJ, Toms River. Beanie Baby Super Show. Holiday Inn, 290 Hwy. 34 E. SH: 9:30am-3:30pm, T: 35. Collectors Showcase, Sandy Grecco, PO Box 189, Atlantic Highlands, NJ 07716. PH: 732-291-1632.

Mar 6 1999 NJ, Colts Neck. Doll & Bear Show. Firehouse #2, Conover Rd. just off Rt. 34. SH: 10am-4pm, A: $1.50, under 6 free. Sallie Natowitz, PO Box 796, Matawan, NJ 07747. PH: 732-583-7915.

Mar 6-7 1999 NJ, Pennsauken. Greenberg's Great Train & Collectible Toy Show. South Jersey Expo Center. SH: Sat. 11am-5pm, Sun. 11am-4pm, A: $5., $2. ages 6-12, under 6 free. Greenberg Shows, Nan Turfle, 7566 Main St., Sykesville, MD 21784. PH: 410-795-7447.

Mar 13 1999 NJ, Hazlet. Beanie Baby Super Show. Hazlet Hotel, 2870 Hwy. 35 (Pky. Exit 117 to Rt. 35 S. 2 mi.). SH: 9:30am-3:30pm, T: 45. Collectors Showcase, Sandy Grecco, PO Box 189, Atlantic Highlands, NJ 07716. PH: 732-291-1632.

Mar 14 1999 NJ, North Bergen. 3rd Annual Toy Soldier, Military Toys & Action Figure Show. Schuetzen Park Ballroom, 32nd St. & Kennedy Blvd. SH: 9am-4pm, T: 100. Ed Gries, PO Box 572, Hackensack, NJ 07602. PH: 201-342-6475.

Mar 14 1999 NJ, Princeton. Doll & Bear Show. Holiday Inn, 4355 Rt. 1 S. at Ridge Rd. SH: 10am-4pm, A: $2., under 6 free. Sallie Natowitz, PO Box 796, Matawan, NJ 07747. PH: 732-583-7915.

Mar 20 1999 NJ, Kenilworth. Beanie Baby Super Show. Kenilworth Inn, off GSP Exit 138, turn L. SH: 9:30am-3:30pm, T: 45. Collectors Showcase, Sandy Grecco, PO Box 189, Atlantic Highlands, NJ 07716. PH: 732-291-1632.

Mar 21 1999 NJ, Mount Laurel. Doll & Bear Show. Radisson Hotel, 915 Rt. 73 at NJ Tpke. Exit 4. SH: 10am-4pm, A: $2., under 6 free. Sallie Natowitz, PO Box 796, Matawan, NJ 07747. PH: 732-583-7915.

Mar 28 1999 NJ, Toms River. Beanie Baby Super Show. Holiday Inn, 290 Hwy. 34 E. SH: 9:30am-3:30pm, T: 35. Collectors Showcase, Sandy Grecco, PO Box 189, Atlantic Highlands, NJ 07716. PH: 732-291-1632.

Apr 17 1999 NJ, Hazlet. Beanie Baby Super Show. Hazlet Hotel, 2870 Hwy. 35 (Pky. Exit 117 to Rt. 35 S. 2 mi.). SH: 9:30am-3:30pm, T: 45. Collectors Showcase, Sandy Grecco, PO Box 189, Atlantic Highlands, NJ 07716. PH: 732-291-1632.

Apr 18 1999 NJ, Kenilworth. Beanie Baby Super Show. Kenilworth Inn, off GSP Exit 138, turn L. SH: 9:30am-3:30pm, T: 45. Collectors Showcase, Sandy Grecco, PO Box 189, Atlantic Highlands, NJ 07716. PH: 732-291-1632.

May 15 1999 NJ, Hazlet. Beanie Baby Super Show. Hazlet Hotel, 2870 Hwy. 35 (Pky. Exit 117 to Rt. 35 S. 2 mi.). SH: 9:30am-3:30pm, T: 45. Collectors Showcase, Sandy Grecco, PO Box 189, Atlantic Highlands, NJ 07716. PH: 732-291-1632.

May 16 1999 NJ, Wayne. The Great Bergen Passaic Toy & Train Show. P.A.L. Hall. SH: 9am-2pm. Vic, PH: 516-653-8133.

May 22 1999 NJ, Kenilworth. Beanie Baby Super Show. Kenilworth Inn, off GSP Exit 138, turn L. SH: 9:30am-3:30pm, T: 45. Collectors Showcase, Sandy Grecco, PO Box 189, Atlantic Highlands, NJ 07716. PH: 732-291-1632.

Jun 12 1999 NJ, Hazlet. Beanie Baby Super Show. Hazlet Hotel, 2870 Hwy. 35 (Pky. Exit 117 to Rt. 35 S. 2 mi.). SH: 9:30am-3:30pm, T: 45. Collectors Showcase, Sandy Grecco, PO Box 189, Atlantic Highlands, NJ 07716. PH: 732-291-1632.

Jun 13 1999 NJ, Hazlet. Collectible Toys, Comic Books, Lineups & Action Figure Show. Hazlet Hotel, 2870 Hwy. 35 (Pky. Exit 117 to Rt. 35 S. 2 mi.). SH: 9:30am-3:30pm, T: 50. Collectors Showcase, Sandy Grecco, PO Box 189, Atlantic Highlands, NJ 07716. PH: 732-291-1632.

Jun 19 1999 NJ, Kenilworth. Beanie Baby Super Show. Kenilworth Inn, off GSP Exit 138, turn L. SH: 9:30am-3:30pm, T: 45. Collectors Showcase, Sandy Grecco, PO Box 189, Atlantic Highlands, NJ 07716. PH: 732-291-1632.

Jul 16-17 1999 NJ, Newark. International Toy Mega Show. Sheraton Airport Hotel, 128 Frontage Rd. SH: 10am-5pm, T: 200. Steve Savino or Ken Laurence, 1020 Arlington Rd., New Millford, NJ 07646. PH: 201-261-8803 or 261-4982.

Sep 12 1999 NJ, Wayne. The Great Bergen Passaic Toy & Train Show. P.A.L. Hall. SH: 9am-2pm. Vic, PH: 516-653-8133.

Dec 12 1999 NJ, Wayne. The Great Bergen Passaic Toy & Train Show. P.A.L. Hall. SH: 9am-2pm. Vic, PH: 516-653-8133.

NEW MEXICO

Feb 21 1999 NM, Albuquerque. Route 66 Antique Toy, Doll & Collectible Show. Hilton, 1901 University Blvd. NE. SH: 9am-3pm, T: 150, A: $2.50, under 10 free. Jim Gallegos, PO Box 15414, Rio Rancho, NM 87174. PH: 505-892-8848.

May 2 1999 NM, Albuquerque. Route 66 Antique Toy, Doll & Collectible Show. Hilton, 1901 University Blvd. NE. SH: 9am-3pm, T: 150, A: $2.50, under 10 free. Jim Gallegos, PO Box 15414, Rio Rancho, NM 87174. PH: 505-892-8848.

NEW YORK

Jan 3 1999 NY, Brooklyn. Toys, Beanie Babies, BB Cards, Comic Books & Coll. Show. Temple Hillel of Flatlands, 2164 Ralph Ave. (at Ave. L). SH: 10am-5pm, T: 50-8'.Scotty O'Donnell, Sunrise Productions, POB 340625, Ryder Station, Brooklyn, NY 11234. PH: 718-251-2075 or 718-241-6477.

Jan 3 1999 NY, Hudson. Philmont Mountain Toy & Railroad Club Toy & Train Swap Meet. American Legion-7 Fairview Ave. SH: 9am-2pm. Rick Washburn, 12 Hudson St., Hudson, NY 12534. PH: 518-828-6508.

Jan 3 1999 NY, Lindenhurst. Northern Spur Train & Diecast, Hess Show. Knights of Columbus Hall, 400 S. Broadway. SH: 8:30am-1pm, T: 105, A: $3., children under 12 free. Carmelo Sancetta, PO Box 1286-M, Bay Shore, NY 11706. PH: 516-666-6855.

Jan 8 1999 NY, Brooklyn. Toys, Beanie Babies, BB Cards, Comic Books & Coll. Show. St. Dominics Church, 20th Ave. & Bayridge Pky. (75th St.). SH: 6pm-10pm, T: 50-9'.Scotty O'Donnell, Sunrise Productions, POB 340625, Ryder Station, Brooklyn, NY 11234. PH: 718-251-2075

or 718-241-6477.

Jan 9 1999 NY, Yonkers. Nostalgia Int'l Toy & Train Show. Yonkers Raceway, Old Glory Bldg. SH: 9am-3pm. George, PH: 518-392-2660.

Jan 10 1999 NY, Elmont-LI. Model Toy, Train, Doll, Miniature & Craft Show. St. Vincent DePaul Parish, 1510 DePaul St. SH: 10am-3pm. Frank Deorio, 1500 DePaul St., Elmont, NY 11003. PH: 516-352-2127.

Jan 17 1999 NY, Bayside. NY Toy Soldier Show. Adria Hotel & Conference Ctr., 221-17 Northern Blvd. SH: 10am-3pm. E. Reinikka, PH: 718-835-9764.

Jan 17 1999 NY, Brooklyn. Toys, Beanie Babies, BB Cards, Comic Books & Coll. Show. Temple Hillel of Flatlands, 2164 Ralph Ave. (at Ave. L). SH: 10am-5pm, T: 50-8'.Scotty O'Donnell, Sunrise Productions, POB 340625, Ryder Station, Brooklyn, NY 11234. PH: 718-251-2075 or 718-241-6477.

Jan 22 1999 NY, Brooklyn. Toys, Beanie Babies, BB Cards, Comic Books & Coll. Show. St. Dominics Church, 20th Ave. & Bayridge Pky. (75th St.). SH: 6pm-10pm, T: 50-9'.Scotty O'Donnell, Sunrise Productions, POB 340625, Ryder Station, Brooklyn, NY 11234. PH: 718-251-2075 or 718-241-6477.

Jan 31 1999 NY, Brooklyn. Toys, Beanie Babies, BB Cards, Comic Books & Coll. Show. Temple Hillel of Flatlands, 2164 Ralph Ave. (at Ave. L). SH: 10am-5pm, T: 50-8'.Scotty O'Donnell, Sunrise Productions, POB 340625, Ryder Station, Brooklyn, NY 11234. PH: 718-251-2075 or 718-241-6477.

Jan 31 1999 NY, Freeport-LI. Toy Memories Antique Toy Show. Recreation Ctr., 130 E. Merrick Rd. SH: 10am-3pm, T: 160. Guy De Marco, PO Box 224, W. Hempstead, NY 11552. PH: 516-593-8198.

Feb 5 1999 NY, Brooklyn. Toys, Beanie Babies, BB Cards, Comic Books & Coll. Show. St. Dominics Church, 20th Ave. & Bayridge Pky. (75th St.). SH: 6pm-10pm, T: 50-9'.Scotty O'Donnell, Sunrise Productions, POB 340625, Ryder Station, Brooklyn, NY 11234. PH: 718-251-2075 or 718-241-6477.

Feb 7 1999 NY, Syracuse. Collectorsfest Show. Fairgrounds, Horticutural Bldg., State Fair Blvd., Rt. 690, Exit 7, Gate 2. SH: 10am-4pm, T: 175, A: $3., under 10 free. Central NY Promos., Lyn Lake, 35 Hubbard St. Apt 1, Cortland, NY 13045. PH: 607-753-8580.

Feb 14 1999 NY, Brooklyn. Toys, Beanie Babies, BB Cards, Comic Books & Coll. Show. Temple Hillel of Flatlands, 2164 Ralph Ave. (at Ave. L). SH: 10am-5pm, T: 50-8'.Scotty O'Donnell, Sunrise Productions, POB 340625, Ryder Station, Brooklyn, NY 11234. PH: 718-251-2075 or 718-241-6477.

Feb 14 1999 NY, Franklin Square. Model Train & Toy Show. VFW Hall, 68 Lincoln Rd. SH: 9am-1pm. Rae Romano, PH: 516-775-4801.

Feb 19 1999 NY, Brooklyn. Toys, Beanie Babies, BB Cards, Comic Books & Coll. Show. St. Dominics Church, 20th Ave. & Bayridge Pky. (75th St.). SH: 6pm-10pm, T: 50-9'.Scotty O'Donnell, Sunrise Productions, POB 340625, Ryder Station, Brooklyn, NY 11234. PH: 718-251-2075 or 718-241-6477.

Feb 21 1999 NY, White Plains. The Great Westchester Toy & Train Show. County Center. SH: 9am-3pm. George, PH: 518-392-2660.

Feb 21 1999 NY, Lindenhurst. Northern Spur Train & Diecast, Hess Show. Knights of Columbus Hall, 400 S. Broadway. SH: 8:30am-1pm, T: 105, A: $3., children under 12 free. Carmelo Sancetta, PO Box 1286-M, Bay Shore, NY 11706. PH: 516-666-6855.

Feb 28 1999 NY, Hauppauge. Long Island Antique & Collectible Toy Expo. Wyndham Wind Watch, 1717 Motor Pky. SH: 10am-3m, T: 123. Toy Box Prods., Jeff Tuthill, PO Box 356, Shirley, NY 11967. PH: 516-395-1899.

Feb 28 1999 NY, Brooklyn. Toys, Beanie Babies, BB Cards, Comic Books & Coll. Show. Temple Hillel of Flatlands, 2164 Ralph Ave. (at Ave. L). SH: 10am-5pm, T: 50-8'.Scotty O'Donnell, Sunrise Productions, POB 340625, Ryder Station, Brooklyn, NY 11234. PH: 718-251-2075 or 718-241-6477.

Mar 5 1999 NY, Brooklyn. Toys, Beanie Babies, BB Cards, Comic Books & Coll. Show. St. Dominics Church, 20th Ave. & Bayridge Pky. (75th St.). SH: 6pm-10pm, T: 50-9'.Scotty O'Donnell, Sunrise Productions, POB 340625, Ryder Station, Brooklyn, NY 11234. PH: 718-251-2075 or 718-241-6477.

Mar 6 1999 NY, Albany. 27th Annual Toy Show. Polish Community Center, Washington Ave. Ext. SH: 9am-2pm. Northland Toy Club, PO Box 2, Medusa, NY 12120. PH: 518-966-5239 eves.

Mar 13-14 1999 NY, Stony Brook. Greenberg's Great Train & Collectible Toy Show. Stony Brook University. SH: Sat. 11am-5pm, Sun. 11am-4pm, A: $5., $2. ages 6-12, under 6 free. Greenberg Shows, Nan Turfle, 7566 Main St., Sykesville, MD 21784. PH: 410-795-7447.

Mar 14 1999 NY, Brooklyn. Toys, Beanie Babies, BB Cards, Comic Books & Coll. Show. Temple Hillel of Flatlands, 2164 Ralph Ave. (at Ave. L). SH: 10am-5pm, T: 50-8'.Scotty O'Donnell, Sunrise Productions, POB 340625, Ryder Station, Brooklyn, NY 11234. PH: 718-251-2075 or 718-241-6477.

Mar 19 1999 NY, Brooklyn. Toys, Beanie Babies, BB Cards, Comic Books & Coll. Show. St. Dominics Church, 20th Ave. & Bayridge Pky. (75th St.). SH: 6pm-10pm, T: 50-9'.Scotty O'Donnell, Sunrise Productions, POB 340625, Ryder Station, Brooklyn, NY 11234. PH: 718-251-2075 or 718-241-6477.

Mar 21 1999 NY, Lindenhurst. Northern Spur Train & Diecast, Hess Show. Knights of Columbus Hall, 400 S. Broadway. SH: 8:30am-1pm, T: 105, A: $3., children under 12 free. Carmelo Sancetta, PO Box 1286-M, Bay Shore, NY 11706. PH: 516-666-6855.

Mar 28 1999 NY, Brooklyn. Toys, Beanie Babies, BB Cards, Comic Books & Coll. Show. Temple Hillel of Flatlands, 2164 Ralph Ave. (at Ave. L). SH: 10am-5pm, T: 50-8'.Scotty O'Donnell, Sunrise Productions, POB 340625, Ryder Station, Brooklyn, NY 11234. PH: 718-251-2075 or 718-241-6477.

Mar 28 1999 NY, Franklin Square. Model Train & Toy Show. VFW Hall, 68 Lincoln Rd. SH: 9am-1pm. Rae Romano, PH: 516-775-4801.

Apr 3 1999 NY, Farmingdale. RepLIcon Show. Polytechnic University Gymnasium (Polytech), Rt. 110 adjacent Republic Airport. SH: 9am-4:30pm. Bruce Drummond, 6 Oakcrest Ct., E. Northport, NY 11731. PH: 516-754-1918.

Apr 9 1999 NY, Brooklyn. Toys, Beanie Babies, BB Cards, Comic Books & Coll. Show. St. Dominics Church, 20th Ave. & Bayridge Pky. (75th St.). SH: 6pm-10pm, T: 50-9'.Scotty O'Donnell, Sunrise Productions, POB 340625, Ryder Station, Brooklyn, NY 11234. PH: 718-251-2075 or 718-241-6477.

Apr 10 1999 NY, Syracuse. Doll, Teddy Bear & Beanie Baby Show. State Fairgrounds, Horticutural Bldg., State Fair Blvd., Rt. 690, Exit 7, Gate 2. SH: 10am-4pm, T: 150, A: $3., under 10 free. Central NY Promos., Lyn Lake, 35 Hubbard St. Apt 1, Cortland, NY 13045. PH: 607-753-8580.

Apr 11 1999 NY, Elmont-LI. Model Toy, Train, Doll, Miniature & Craft Show. St. Vincent DePaul Parish, 1510 DePaul St. SH: 10am-3pm. Frank Deorio, 1500 DePaul St., Elmont, NY 11003. PH: 516-352-2127.

Apr 11 1999 NY, Syracuse. Collectorsfest Show. Fairgrounds, Horticutural Bldg., State Fair Blvd., Rt. 690, Exit 7, Gate 2. SH: 10am-4pm, T: 175, A: $3., under 10 free. Central NY Promos., Lyn Lake, 35 Hubbard St. Apt 1, Cortland, NY 13045. PH: 607-753-8580.

Apr 17 1999 NY, New Hartford. Heritage Doll Club Show. First United Methodist Church, 105 Genesee St. SH: 10am-4pm, A: $3.50. Mary Polera, 1236 Hammond Ave., Utica, NY 13501. PH: 315-735-5628.

Apr 18 1999 NY, Brooklyn. Toys, Beanie Babies, BB Cards, Comic Books & Coll. Show. Temple Hillel of Flatlands, 2164 Ralph Ave. (at Ave. L). SH: 10am-5pm, T: 50-8'.Scotty O'Donnell, Sunrise Productions, POB 340625, Ryder Station, Brooklyn, NY 11234. PH: 718-251-2075 or 718-241-6477.

Apr 23 1999 NY, Brooklyn. Toys, Beanie Babies, BB Cards, Comic Books & Coll. Show. St. Dominics Church, 20th Ave. & Bayridge Pky. (75th St.). SH: 6pm-10pm, T: 50-9'.Scotty O'Donnell, Sunrise Productions, POB 340625, Ryder Station, Brooklyn, NY 11234. PH: 718-251-2075 or 718-241-6477.

Apr 24 1999 NY, New York. The Joe & Marl Show. Holiday Inn Rockville Centre, 173 Sunrise Hwy. SH: 10am-3pm, A: $5., $2. under 12. Marl, PH: 941-751-6275 or Joe, PH: 213-953-6490.

Apr 25 1999 NY, White Plains. The Great Westchester Toy & Train Show. County Center. SH: 9am-3pm. George, PH: 518-392-2660.

Apr 25 1999 NY, Brooklyn. Toys, Beanie Babies, BB Cards, Comic Books & Coll. Show. Temple Hillel of Flatlands, 2164 Ralph Ave. (at Ave. L). SH: 10am-5pm, T: 50-8'.Scotty O'Donnell, Sunrise Productions, POB 340625, Ryder Station, Brooklyn, NY 11234. PH: 718-251-2075 or 718-241-6477.

Apr 30 1999 NY, Brooklyn. Toys, Beanie Babies, BB Cards, Comic Books & Coll. Show. St. Dominics Church, 20th Ave. & Bayridge Pky. (75th St.). SH: 6pm-10pm, T: 50-9'.Scotty O'Donnell, Sunrise Productions, POB 340625, Ryder Station, Brooklyn, NY 11234. PH: 718-251-2075 or 718-241-6477.

May 9 1999 NY, Brooklyn. Toys, Beanie Babies, BB Cards, Comic Books & Coll. Show. Temple Hillel of Flatlands, 2164 Ralph Ave. (at Ave. L). SH: 10am-5pm, T: 50-8'.Scotty O'Donnell, Sunrise Productions, POB 340625, Ryder Station, Brooklyn, NY 11234. PH: 718-251-2075 or 718-241-6477.

May 21 1999 NY, Brooklyn. Toys, Beanie Babies, BB Cards, Comic Books & Coll. Show. St. Dominics Church, 20th Ave. & Bayridge Pky. (75th St.). SH: 6pm-10pm, T: 50-9'.Scotty O'Donnell, Sunrise Productions, POB 340625, Ryder Station, Brooklyn, NY 11234. PH: 718-251-2075 or 718-241-6477.

May 23 1999 NY, Brooklyn. Toys, Beanie Babies, BB Cards, Comic Books & Coll. Show. Temple Hillel of Flatlands, 2164 Ralph Ave. (at Ave. L). SH: 10am-5pm, T: 50-8'.Scotty O'Donnell, Sunrise Productions, POB 340625, Ryder Station, Brooklyn, NY 11234. PH: 718-251-2075 or 718-241-6477.

May 23 1999 NY, Franklin Square. Model Train & Toy Show. VFW Hall, 68 Lincoln Rd. SH: 9am-1pm. Rae Romano, PH: 516-775-4801.

May 28 1999 NY, Brooklyn. Toys, Beanie Babies, BB Cards, Comic Books & Coll. Show. St. Dominics Church, 20th Ave. & Bayridge Pky. (75th St.). SH: 6pm-10pm, T: 50-9'.Scotty O'Donnell, Sunrise Productions, POB 340625, Ryder Station, Brooklyn, NY 11234. PH: 718-251-2075 or 718-241-6477.

Jun 6 1999 NY, Brooklyn. Toys, Beanie Babies, BB Cards, Comic Books & Coll. Show. Temple Hillel of Flatlands, 2164 Ralph Ave. (at Ave. L). SH: 10am-5pm, T: 50-8'.Scotty O'Donnell, Sunrise Productions, POB 340625, Ryder Station, Brooklyn, NY 11234. PH: 718-251-2075 or 718-241-6477.

Jun 6 1999 NY, Lindenhurst. Northern Spur Train & Diecast, Hess Show. Knights of Columbus Hall, 400 S. Broadway. SH: 8:30am-1pm, T: 105, A: $3., children under 12 free. Carmelo Sancetta, PO Box 1286-M, Bay Shore, NY 11706. PH: 516-666-6855.

Jun 11 1999 NY, Brooklyn. Toys, Beanie Babies, BB Cards, Comic Books & Coll. Show. St. Dominics Church, 20th Ave. & Bayridge Pky. (75th St.). SH: 6pm-10pm, T: 50-9'.Scotty O'Donnell, Sunrise Productions, POB 340625, Ryder Station, Brooklyn, NY 11234. PH: 718-251-2075 or 718-241-6477.

Jun 20 1999 NY, Brooklyn. Toys, Beanie Babies, BB Cards, Comic Books & Coll. Show. Temple Hillel of Flatlands, 2164 Ralph Ave. (at Ave. L). SH: 10am-5pm, T: 50-8'.Scotty O'Donnell, Sunrise Productions, POB 340625, Ryder Station, Brooklyn, NY 11234. PH: 718-251-2075 or 718-241-6477.

Jun 25 1999 NY, Brooklyn. Toys, Beanie Babies, BB Cards, Comic Books & Coll. Show. St. Dominics Church, 20th Ave. & Bayridge Pky. (75th St.). SH: 6pm-10pm, T: 50-9'.Scotty O'Donnell, Sunrise Productions, POB 340625, Ryder Station, Brooklyn, NY 11234. PH: 718-251-2075 or 718-241-6477.

Jun 27 1999 NY, Franklin Square. Model Train & Toy Show. VFW Hall, 68 Lincoln Rd. SH: 9am-1pm. Rae Romano, PH: 516-775-4801.

Aug 1 1999 NY, Lindenhurst. Northern Spur Train & Diecast, Hess Show. Knights of Columbus Hall, 400 S. Broadway. SH: 8:30am-1pm, T: 105, A: $3., children under 12 free. Carmelo Sancetta, PO Box 1286-M, Bay Shore, NY 11706. PH: 516-666-6855.

Aug 8 1999 NY, Elmont-LI. Model Toy, Train, Doll, Miniature & Craft Show. St. Vincent DePaul Parish, 1510 DePaul St. SH: 10am-3pm. Frank Deorio, 1500 DePaul St., Elmont, NY 11003. PH: 516-352-2127.

Aug 15 1999 NY, Franklin Square. Model Train & Toy Show. VFW Hall, 68 Lincoln Rd. SH: 9am-1pm. Rae Romano, PH: 516-775-4801.

Sep 19 1999 NY, Franklin Square. Model Train & Toy Show. VFW Hall, 68 Lincoln Rd. SH: 9am-1pm. Rae Romano, PH: 516-775-4801.

Oct 30 1999 NY, Syracuse. Doll, Teddy Bear & Beanie Baby Show. State Fairgrounds, Horticutural Bldg., State Fair Blvd., Rt. 690, Exit 7, Gate 2. SH: 10am-4pm, T: 150, A: $3., under 10 free. Central NY Promos., Lyn Lake, 35 Hubbard St. Apt 1, Cortland, NY 13045. PH: 607-753-8580.

Oct 31 1999 NY, Elmont-LI. Model Toy, Train, Doll, Miniature & Craft Show. St. Vincent DePaul Parish, 1510 DePaul St. SH: 10am-3pm. Frank Deorio, 1500 DePaul St., Elmont, NY 11003. PH: 516-352-2127.

Oct 31 1999 NY, Syracuse. Collectorsfest Show. Fairgrounds, Horticutural Bldg., State Fair Blvd., Rt. 690, Exit 7, Gate 2. SH: 10am-4pm, T: 175, A: $3., under 10 free. Central NY Promos., Lyn Lake, 35 Hubbard St. Apt 1, Cortland, NY 13045. PH: 607-753-8580.

Nov 7 1999 NY, Lindenhurst. Northern Spur Train & Diecast, Hess Show. Knights of Columbus Hall, 400 S. Broadway. SH: 8:30am-1pm, T: 105, A: $3., children under 12 free. Carmelo Sancetta, PO Box 1286-M, Bay Shore, NY 11706. PH: 516-666-6855.

Nov 14 1999 NY, Franklin Square. Model Train & Toy Show. VFW Hall, 68 Lincoln Rd. SH: 9am-1pm. Rae Romano, PH: 516-775-4801.

Dec 5 1999 NY, Elmont-LI. Model Toy, Train, Doll, Miniature & Craft Show. St. Vincent DePaul Parish, 1510 DePaul St. SH: 10am-3pm. Frank Deorio, 1500 DePaul St., Elmont, NY 11003. PH: 516-352-2127.

Dec 19 1999 NY, White Plains. The Great Westchester Toy & Train Show. County Center. SH: 9am-3pm. George, PH: 518-392-2660.

Dec 26 1999 NY, Franklin Square. Model Train & Toy Show. VFW Hall, 68 Lincoln Rd. SH: 9am-1pm. Rae Romano, PH: 516-775-4801.

NORTH CAROLINA

Jan 16-17 1999 NC, Raleigh. 30th North State Toy Collectors Show. State Fairgrounds, 1025 Blueridge Blvd. SH: Sat. 9am-5pm, Sun. 10am-4:30pm, T: 285. Carolina Hobby Expo, PH: 704-786-8373.

Jan 23 1999 NC, Concord. 4th Annual Toy Vehicle Show. Nat'l. Guard Armory, Hwy. 49 & Old Charlotte Rd. SH: 9am-4pm, T: 105. Carolina Hobby Expo, PH: 704-786-8373.

Jan 24 1999 NC, Concord. 2nd Annual Barbie Doll & Beanie Babies Show. Nat'l. Guard Armory, Hwy. 49 & Old Charlotte Rd. SH: 10am-4pm, T: 105. Carolina Hobby Expo, PH: 704-786-8373.

Jan 30-31 1999 NC, Charlotte. Toy, Train & Doll Show. Metrolina Expo Center, I-77, Exit 16A. SH: Sat. 9am-5pm, Sun. 10am-4pm, T: 500, A: $5., under 12 free. Tri-City Shows, PO Box 825, Johnson City, TN 37605. PH/FAX: 888-955-TOYS.

Jan 30 1999 NC, Fayetteville. 27th Sandhills Toy & Hobby Show. Charlie Rose Expo Ctr., just off Hwy. 301 S. SH: 9am-4pm, T: 225. Carolina Hobby Expo, PH: 704-786-8373.

Feb 7 1999 NC, Lake Norman-Cornelius. Beanie Babies & Barbie Doll Show. Holiday Inn, I-77, Exit 28. SH: 10am-4pm, T: 60. JJ's Toys, PH: 704-662-8155.

Oct 23-24 1999 NC, Charlotte. Toy, Train & Doll Show. Metrolina Expo Center, I-77, Exit 16A. SH: Sat. 9am-5pm, Sun. 10am-4pm, T: 500, A: $5., under 12 free. Tri-City Shows, PO Box 825, Johnson City, TN 37605. PH/FAX: 888-955-TOYS.

OHIO

Jan 2 1999 OH, Middletown. Beanie Baby Show. Garden Inn & Suites, I-75 to Exit 32, Rt. 122. SH: 10am-4pm, T: 22. The Strike Zone, PH: 513-727-9378.

Jan 3 1999 OH, Columbus. Book & Paper Fair. Veterans Memorial Hall, 300 W. Broad St. SH: 10am-5pm, T: 325. Columbus Prods., Inc., PO Box 261016, Columbus, OH 43226. PH: 614-781-0070.

Jan 5 1999 OH, Toledo. Every Tues. Collectibles Show. St. Clements Hall, 2990 Tremainsville Rd. SH: 3:30pm-8pm, T: 50-8', A: free. Dan Freiheit, PO Box 8404, Toledo, OH 43623. PH: 419-843-7171.

Jan 9-10 1999 OH, Columbus. Greenberg's Great Train & Collectible Toy Show. Franklin Cty. Veterans Memorial, 300 W. Broad St. SH: Sat. 11am-5pm, Sun. 11am-4pm, A: $5., $2. ages 6-12, under 6 free. Greenberg Shows, Nan Turfle, 7566 Main St., Sykesville, MD 21784. PH: 410-795-7447.

Jan 9 1999 OH, Celina. Hospice Race & Memorabilia Show. Knights of Columbus, 129 N. Vine St. SH: 9am-4pm, T: 35, A: $1. Marge or Dave Schwartz, 202 S. Main St., Celina, OH 45822. PH: 419-586-5385.

Jan 10 1999 OH, Massillon. Winter Train & Toy Show. K of C Hall, 988 Cherry Rd. NW. SH: 10am-4pm, T: 8'. Glenn Groff, 3385 Brigadoon Circle SW, Dalton, OH 44618. PH: 800-833-0612 after 6pm.

Jan 12 1999 OH, Toledo. Every Tues. Collectibles Show. St. Clements Hall, 2990 Tremainsville Rd. SH: 3:30pm-8pm, T: 50-8', A: free. Dan Freiheit, PO Box 8404, Toledo, OH 43623. PH: 419-843-7171.

Jan 17 1999 OH, Parma Hts. Cleveland Nat'l. Beanie Baby Show. Valley Forge H.S., 9999 Independence Blvd. SH: 9am-3pm. Bob Frieden, 9695 Chillicothe Rd., Kirtland, OH 44094. PH: 440-256-8141.

Jan 17 1999 OH, Richfield-Cleveland. Winter Train & Toy

Show. Holiday Inn, 4742 Brecksville Rd. SH: 10am-4pm, T: 8'. Glenn Groff, 3385 Brigadoon Circle SW, Dalton, OH 44618. PH: 800-833-0612.

Jan 19 1999 OH, Toledo. Every Tues. Collectibles Show. St. Clements Hall, 2990 Tremainsville Rd. SH: 3:30pm-8pm, T: 50-8', A: free. Dan Freiheit, PO Box 8404, Toledo, OH 43623. PH: 419-843-7171.

Jan 26 1999 OH, Toledo. Every Tues. Collectibles Show. St. Clements Hall, 2990 Tremainsville Rd. SH: 3:30pm-8pm, T: 50-8', A: free. Dan Freiheit, PO Box 8404, Toledo, OH 43623. PH: 419-843-7171.

Jan 30-31 1999 OH, New Philadelphia. Winter Train & Toy Show. New Towne Mall, 410 E. Mill St. SH: Sat. 11am-9pm, Sun. 11am-6pm, T: 8', A: $3., 12 & under free. Glenn Groff, 3385 Brigadoon Circle SW, Dalton, OH 44618. PH: 800-833-0612.

Jan 31 1999 OH, Mansfield. Toy & Collectible Show. Richland Co. Fairgrounds, Trimble Rd. Exit off US Rt. 30. SH: 10am-4pm, T: 175. Kevin Spore, PO Box 9014, Lexington, OH 44904. PH: 419-756-3904.

Jan 31 1999 OH, Cleveland-Brunswick. Winter Train & Toy Show. Sally's Party Ctr., 1480 Pearl Rd. SH: 10am-4pm, T: 8'. Glenn Groff, 3385 Brigadoon Circle SW, Dalton, OH 44618. PH: 800-833-0612.

Feb 2 1999 OH, Toledo. Every Tues. Collectibles Show. St. Clements Hall, 2990 Tremainsville Rd. SH: 3:30pm-8pm, T: 50-8', A: free. Dan Freiheit, PO Box 8404, Toledo, OH 43623. PH: 419-843-7171.

Feb 9 1999 OH, Toledo. Every Tues. Collectibles Show. St. Clements Hall, 2990 Tremainsville Rd. SH: 3:30pm-8pm, T: 50-8', A: free. Dan Freiheit, PO Box 8404, Toledo, OH 43623. PH: 419-843-7171.

Feb 13-14 1999 OH, Cincinnati. Beanies Baby & Toy Expo. Convention Ctr. T: 8'. CEI Sports, Charles Sotto, PH: 513-936-9941.

Feb 14 1999 OH, Akron. Winter Train & Toy Show. Tadmor Temple Hall, 3000 Krebs Dr. SH: 10am-4pm, T: 8'. Glenn Groff, 3385 Brigadoon Circle SW, Dalton, OH 44618. PH: 800-833-0612.

Feb 16 1999 OH, Toledo. Every Tues. Collectibles Show. St. Clements Hall, 2990 Tremainsville Rd. SH: 3:30pm-8pm, T: 50-8', A: free. Dan Freiheit, PO Box 8404, Toledo, OH 43623. PH: 419-843-7171.

Feb 20-21 1999 OH, Niles. Greenberg's Great Train & Collectible Toy Show. Eastwood Expo Center, 5555 Youngstown-Warren Rd. SH: Sat. 11am-5pm, Sun. 11am-4pm, A: $5., $2. ages 6-12, under 6 free. Greenberg Shows, Nan Turfle, 7566 Main St., Sykesville, MD 21784. PH: 410-795-7447.

Feb 23 1999 OH, Toledo. Every Tues. Collectibles Show. St. Clements Hall, 2990 Tremainsville Rd. SH: 3:30pm-8pm, T: 50-8', A: free. Dan Freiheit, PO Box 8404, Toledo, OH 43623. PH: 419-843-7171.

Feb 28 1999 OH, Akron-Fairlawn. Winter Train & Toy Show. Hilton Inn West, 3180 W. Market. SH: 10am-4pm, T: 8'. Glenn Groff, 3385 Brigadoon Circle SW, Dalton, OH 44618. PH: 800-833-0612.

Mar 2 1999 OH, Toledo. Every Tues. Collectibles Show. St. Clements Hall, 2990 Tremainsville Rd. SH: 3:30pm-8pm, T: 50-8', A: free. Dan Freiheit, PO Box 8404, Toledo, OH 43623. PH: 419-843-7171.

Mar 7 1999 OH, Westlake. Cleveland Toy Show. Holiday Inn, 1100 Crocker Rd. (I-90 W. & Crocker Bassett Rd. Exit). SH: 9am-2pm. Bob Frieden, 9695 Chillicothe Rd., Kirtland, OH 44094. PH: 440-256-8141.

Mar 7 1999 OH, Marietta. Winter Train & Toy Show. LaFayette Hotel, 101 Front St. SH: 10am-4pm, T: 8'. Glenn Groff, 3385 Brigadoon Circle SW, Dalton, OH 44618. PH: 800-833-0612.

Mar 7 1999 OH, Columbus. Annual Doll Show. Aladdin Temple, 2850 Stelzer Rd. SH: 10am-4pm, A: $3., $1.50. Vivian Ashbaugh, Box 468, Pataskala, OH 43062. PH: 740-587-4722.

Mar 9 1999 OH, Toledo. Every Tues. Collectibles Show. St. Clements Hall, 2990 Tremainsville Rd. SH: 3:30pm-8pm, T: 50-8', A: free. Dan Freiheit, PO Box 8404, Toledo, OH 43623. PH: 419-843-7171.

Mar 14 1999 OH, Toledo. Book & Paper Fair. SeaGate Centre, 401 Jefferson Ave. Columbus Prods., Inc., PO Box 261016, Columbus, OH 43226. PH: 614-781-0070.

Mar 14 1999 OH, Findlay. Putnam Assoc. of Railfans Model Railroad & Farm Toy Show & Swap Meet. High School, 1200 Broad Ave. & US 224. SH: 10am-4pm, A: $3., under 12 free. Randy Gratz, 3396 Old St. Rt. 224, Ottawa, OH 45875. PH: 419-456-3325.

Mar 16 1999 OH, Toledo. Every Tues. Collectibles Show. St. Clements Hall, 2990 Tremainsville Rd. SH: 3:30pm-8pm, T: 50-8', A: free. Dan Freiheit, PO Box 8404, Toledo, OH 43623. PH: 419-843-7171.

Mar 21 1999 OH, Mansfield. Toy & Collectible Show. Richland Co. Fairgrounds, Trimble Rd. Exit off US Rt. 30. SH: 10am-4pm, T: 175. Kevin Spore, PO Box 9014, Lexington, OH 44904. PH: 419-756-3904.

Mar 23 1999 OH, Toledo. Every Tues. Collectibles Show. St. Clements Hall, 2990 Tremainsville Rd. SH: 3:30pm-8pm, T: 50-8', A: free. Dan Freiheit, PO Box 8404, Toledo, OH 43623. PH: 419-843-7171.

Mar 30 1999 OH, Toledo. Every Tues. Collectibles Show. St. Clements Hall, 2990 Tremainsville Rd. SH: 3:30pm-8pm, T: 50-8', A: free. Dan Freiheit, PO Box 8404, Toledo, OH 43623. PH: 419-843-7171.

Apr 6 1999 OH, Toledo. Every Tues. Collectibles Show. St. Clements Hall, 2990 Tremainsville Rd. SH: 3:30pm-8pm, T: 50-8', A: free. Dan Freiheit, PO Box 8404, Toledo, OH 43623. PH: 419-843-7171.

Apr 13 1999 OH, Toledo. Every Tues. Collectibles Show. St. Clements Hall, 2990 Tremainsville Rd. SH: 3:30pm-8pm, T: 50-8', A: free. Dan Freiheit, PO Box 8404, Toledo, OH 43623. PH: 419-843-7171.

Apr 18 1999 OH, Boston Hts. Cleveland Nat'l. Beanie Baby Show. Hudson Holiday Inn, 240 Hines Hill Rd. (Rt. 8 & Ohio Tpke. Exit 12). SH: 9am-3pm. Bob Frieden, 9695 Chillicothe Rd., Kirtland, OH 44094. PH: 440-256-8141.

Apr 20 1999 OH, Toledo. Every Tues. Collectibles Show. St. Clements Hall, 2990 Tremainsville Rd. SH: 3:30pm-8pm, T: 50-8', A: free. Dan Freiheit, PO Box 8404, Toledo, OH 43623. PH: 419-843-7171.

Apr 27 1999 OH, Toledo. Every Tues. Collectibles Show. St. Clements Hall, 2990 Tremainsville Rd. SH: 3:30pm-8pm, T: 50-8', A: free. Dan Freiheit, PO Box 8404, Toledo, OH 43623. PH: 419-843-7171.

May 4 1999 OH, Toledo. Every Tues. Collectibles Show. St. Clements Hall, 2990 Tremainsville Rd. SH: 3:30pm-8pm, T: 50-8', A: free. Dan Freiheit, PO Box 8404, Toledo, OH 43623. PH: 419-843-7171.

May 11 1999 OH, Toledo. Every Tues. Collectibles Show. St. Clements Hall, 2990 Tremainsville Rd. SH: 3:30pm-8pm, T: 50-8', A: free. Dan Freiheit, PO Box 8404, Toledo, OH 43623. PH: 419-843-7171.

May 16 1999 OH, Columbus. Book & Paper Fair. Veterans Memorial Hall, 300 W. Broad St. SH: 10am-5pm, T: 325. Columbus Prods., Inc., PO Box 261016, Columbus, OH 43226. PH: 614-781-0070.

May 18 1999 OH, Toledo. Every Tues. Collectibles Show. St. Clements Hall, 2990 Tremainsville Rd. SH: 3:30pm-8pm, T: 50-8', A: free. Dan Freiheit, PO Box 8404, Toledo, OH 43623. PH: 419-843-7171.

May 25 1999 OH, Toledo. Every Tues. Collectibles Show. St. Clements Hall, 2990 Tremainsville Rd. SH: 3:30pm-8pm, T: 50-8', A: free. Dan Freiheit, PO Box 8404, Toledo, OH 43623. PH: 419-843-7171.

Jun 1 1999 OH, Toledo. Every Tues. Collectibles Show. St. Clements Hall, 2990 Tremainsville Rd. SH: 3:30pm-8pm, T: 50-8', A: free. Dan Freiheit, PO Box 8404, Toledo, OH 43623. PH: 419-843-7171.

Jun 5 1999 OH, Delaware. Hot Wheels, Racing & Collector Toy Show. Fairgrounds, Jr. Fair Bldg., Pennsylvania Ave. SH: 9am-3pm, T: 65. Jack Seitter, 85 Westland Way, Delaware, OH 43015. PH: 740-362-8652.

Jun 6 1999 OH, Delaware. 9th Annual Farm Toy Show. Fairgrounds, Jr. Fair Bldg., Pennsylvania Ave. SH: 9am-3pm, T: 65. Jack Seitter, 85 Westland Way, Delaware, OH 43015. PH: 740-362-8652.

Jun 8 1999 OH, Toledo. Every Tues. Collectibles Show. St. Clements Hall, 2990 Tremainsville Rd. SH: 3:30pm-8pm, T: 50-8', A: free. Dan Freiheit, PO Box 8404, Toledo, OH 43623. PH: 419-843-7171.

Jun 15 1999 OH, Toledo. Every Tues. Collectibles Show. St. Clements Hall, 2990 Tremainsville Rd. SH: 3:30pm-8pm, T: 50-8', A: free. Dan Freiheit, PO Box 8404, Toledo, OH 43623. PH: 419-843-7171.

Jun 22 1999 OH, Toledo. Every Tues. Collectibles Show. St. Clements Hall, 2990 Tremainsville Rd. SH: 3:30pm-8pm, T: 50-8', A: free. Dan Freiheit, PO Box 8404, Toledo, OH 43623. PH: 419-843-7171.

Jun 26 1999 OH, Westlake. The Joe & Marl Show. Holiday Inn, 1100 Crocker Rd. SH: 10am-3pm, A: $5., $2. under 12. Marl, PH: 941-751-6275 or Joe, PH: 213-953-6490.

Jun 29 1999 OH, Toledo. Every Tues. Collectibles Show. St. Clements Hall, 2990 Tremainsville Rd. SH: 3:30pm-8pm, T: 50-8', A: free. Dan Freiheit, PO Box 8404, Toledo, OH 43623. PH: 419-843-7171.

Sep 12 1999 OH, Columbus. Book & Paper Fair. Veterans Memorial Hall, 300 W. Broad St. SH: 10am-5pm, T: 325. Columbus Prods., Inc., PO Box 261016, Columbus, OH 43226. PH: 614-781-0070.

OKLAHOMA

Feb 20 1999 OK, Tulsa. Toy Car & Memorabilia Collectibles Show. Nat'l. Guard Bldg., 3901 E. 15th (N. of Fairgrounds). SH: 9am-4pm, T: 100. Randy Smith, 8335 E. 14th, Tulsa, OK 74112. PH: 918-835-6074.

PENNSYLVANIA

Jan 3 1999 PA, Gilbertsville. Train-O-Rama & Toy Show. Fire House, Rt. 73 (E. of Rt. 100). SH: 9am-2pm. Mary Preudhomme, 233 Long Lane Rd., Boyertown, PA 19512. PH: 610-367-7857.

Jan 10 1999 PA, New Hope. Toy & Train Show. Eagle Fire Co., Rt. 202 & Sugan Rd. SH: 8am-1pm, T: 100. Fred Dauncey, PO Box 222, S. Plainfield, NJ 07080. PH: 908-755-7989.

Jan 16-17 1999 PA, Lebanon. Greenberg's Great Train & Collectible Toy Show. Lebanon Valley Expo Ctr., PA Tpke. Exit 20, Rt. 7 N. to Rocherty Rd. SH: Sat. 11am-5pm, Sun. 11am-4pm, A: $5., $2. ages 6-12, under 6 free. Greenberg Shows, Nan Turfle, 7566 Main St., Sykesville, MD 21784. PH: 410-795-7447.

Jan 24 1999 PA, Mars-N. Pittsburgh. Winter Train & Toy Show. Sheraton Inn North, 910 Sheraton Dr. SH: 10am-4pm, T: 8'. Glenn Groff, 3385 Brigadoon Circle SW, Dalton, OH 44618. PH: 800-833-0612 after 6pm.

Jan 31 1999 PA, Shrewsbury. Super Sunday Collectors Toy Show. Fire Hall. SH: 9am-1:30pm, T: 6'. Joe Golabiewski, PH: 410-592-5854 or Carl Daehnke, PH: 717-764-5411.

Feb 6-7 1999 PA, Ft. Washington. Greenberg's Great Train & Collectible Toy Show. Expo Ctr., PA Tpke., Exit 26. SH: Sat. 11am-5pm, Sun. 11am-4pm, A: $5., $2. ages 6-12, under 6 free. Greenberg Shows, Nan Turfle, 7566 Main St., Sykesville, MD 21784. PH: 410-795-7447.

Feb 7 1999 PA, New Hope. Toy & Train Show. Eagle Fire Co., Rt. 202 & Sugan Rd. SH: 8am-1pm, T: 100. Fred Dauncey, PO Box 222, S. Plainfield, NJ 07080. PH: 908-755-7989.

Feb 13-14 1999 PA, Monroeville. Greenberg's Great Train & Collectible Toy Show. Pittsburgh ExpoMart, US Bus. Rt. 22. SH: Sat. 11am-5pm, Sun. 11am-4pm, A: $5., $2. ages 6-12, under 6 free. Greenberg Shows, Nan Turfle, 7566 Main St., Sykesville, MD 21784. PH: 410-795-7447.

Feb 14 1999 PA, Gilbertsville. Train-O-Rama & Toy Show. Fire House, Rt. 73 (E. of Rt. 100). SH: 9am-2pm. Mary Preudhomme, 233 Long Lane Rd., Boyertown, PA 19512. PH: 610-367-7857.

Feb 21 1999 PA, Pittsburgh. Winter Train & Toy Show. Holiday Inn South, 164 Fort Couch Rd. at US 19. SH: 10am-4pm, T: 8', A: $3., 12 & under free. Glenn Groff, 3385 Brigadoon Circle SW, Dalton, OH 44618. PH: 800-833-0612.

Feb 27-28 1999 PA, Allentown. ATMA Spring Thaw Train Meet. Agricultural Hall, Fairgrounds, 17th & Chew Sts. SH: 9am-3pm, T: 500. Bob House, 1120 S. Jefferson St., Allentown, PA 18103. PH: 610-821-7886.

Mar 7 1999 PA, New Hope. Toy & Train Show. Eagle Fire Co., Rt. 202 & Sugan Rd. SH: 8am-1pm, T: 100. Fred Dauncey, PO Box 222, S. Plainfield, NJ 07080. PH: 908-755-7989.

Mar 13-15 1999 PA, Philadelphia. East Coast Hobby Show. Ft. Washington Expo Ctr. T: 350 booths. Scott Pressman, PH: 800-252-4757.

Mar 13 1999 PA, Chaddsford. Hillendale's Spring Training Train Show. Hillendale Elem. School, 1850 Hillendale Rd. SH: 9am-2pm, T: 175-6'. Tom Marinelli, 1850 Hillendale Rd., Chaddsford, PA 19317. PH: 610-388-1439.

Apr 11 1999 PA, New Hope. Toy & Train Show. Eagle Fire Co., Rt. 202 & Sugan Rd. SH: 8am-1pm, T: 100. Fred Dauncey, PO Box 222, S. Plainfield, NJ 07080. PH: 908-755-7989.

Apr 24 1999 PA, Allentown. Toy Extravaganza. Merchants Square Mall, S. 12th & Vultee Sts. SH: 9am-6pm. Comic Vault, 4672 Broadway, Tilghman Square, Allentown, PA 18104. PH: 610-395-0979.

May 2 1999 PA, New Hope. Toy & Train Show. Eagle Fire Co., Rt. 202 & Sugan Rd. SH: 8am-1pm, T: 100. Fred Dauncey, PO Box 222, S. Plainfield, NJ 07080. PH: 908-755-7989.

Jun 6 1999 PA, New Hope. Toy & Train Show. Eagle Fire Co., Rt. 202 & Sugan Rd. SH: 8am-1pm, T: 100. Fred Dauncey, PO Box 222, S. Plainfield, NJ 07080. PH: 908-755-7989.

Jun 13 1999 PA, Philadelphia. The Joe & Marl Show. Airport-Marriott Hotel, Arrivals Rd. SH: 10am-3pm, A: $5., $2. under 12. Marl, PH: 941-751-6275 or Joe, PH: 213-953-6490.

Jun 18-20 1999 PA, Pittsburgh. 23rd Annual Parts-A-Rama II Show. Butler Fairgrounds, I-79, Exit 29 to PA Rt. 422 E. PH: 412-366-7154.

Jul 11 1999 PA, New Hope. Toy & Train Show. Eagle Fire Co., Rt. 202 & Sugan Rd. SH: 8am-1pm, T: 100. Fred Dauncey, PO Box 222, S. Plainfield, NJ 07080. PH: 908-755-7989.

Aug 22 1999 PA, Gilbertsville. Train-O-Rama & Toy Show. Fire House, Rt. 73 (E. of Rt. 100). SH: 9am-2pm. Mary Preudhomme, 233 Long Lane Rd., Boyertown, PA 19512. PH: 610-367-7857.

Sep 5 1999 PA, New Hope. Toy & Train Show. Eagle Fire Co., Rt. 202 & Sugan Rd. SH: 8am-1pm, T: 100. Fred Dauncey, PO Box 222, S. Plainfield, NJ 07080. PH: 908-755-7989.

Oct 3 1999 PA, New Hope. Toy & Train Show. Eagle Fire Co., Rt. 202 & Sugan Rd. SH: 8am-1pm, T: 100. Fred Dauncey, PO Box 222, S. Plainfield, NJ 07080. PH: 908-755-7989.

Nov 7 1999 PA, New Hope. Toy & Train Show. Eagle Fire Co., Rt. 202 & Sugan Rd. SH: 8am-1pm, T: 100. Fred Dauncey, PO Box 222, S. Plainfield, NJ 07080. PH: 908-755-7989.

Dec 12 1999 PA, New Hope. Toy & Train Show. Eagle Fire Co., Rt. 202 & Sugan Rd. SH: 8am-1pm, T: 100. Fred Dauncey, PO Box 222, S. Plainfield, NJ 07080. PH: 908-755-7989.

TENNESSEE

Jan 10 1999 TN, Nashville. Hot Wheels Diecast Collectors Show. Super 8 Motel, I-24 & Harding Place. SH: 9am-4pm, T: 40. Bruce Amato, PO Box 2821, Hendersonville, TN 37077. PH: 615-824-1752.

Jan 16-17 1999 TN, Knoxville. Toy, Train, Doll & Hobby Show. Civic Coliseum, I-40, Exit 388-A. SH: Sat. 9am-5pm, Sun. 10am-4pm, T: 400, A: $5., under 12 free. Tri-City Shows, PO Box 825, Johnson City, TN 37605. PH/FAX: 888-955-TOYS.

Jan 23-24 1999 TN, Johnson City-Bristol-Kingsport. Toy, Train & Doll Show. Appalachian Fairgrounds, I-181, Exit 42. SH: Sat. 9am-5pm, Sun. 10am-4pm, T: 300, A: $5., under 12 free. Tri-City Shows, PO Box 825, Johnson City, TN 37605. PH/FAX: 888-955-TOYS.

Mar 7 1999 TN, Nashville. Hot Wheels Diecast Collectors Show. Super 8 Motel, I-24 & Harding Place. SH: 9am-4pm, T: 40. Bruce Amato, PO Box 2821, Hendersonville, TN 37077. PH: 615-824-1752.

May 2 1999 TN, Nashville. Hot Wheels Diecast Collectors Show. Super 8 Motel, I-24 & Harding Place. SH: 9am-4pm, T: 40. Bruce Amato, PO Box 2821, Hendersonville, TN 37077. PH: 615-824-1752.

Jul 11 1999 TN, Nashville. Hot Wheels Diecast Collectors Show. Super 8 Motel, I-24 & Harding Place. SH: 9am-4pm, T: 40. Bruce Amato, PO Box 2821, Hendersonville, TN 37077. PH: 615-824-1752.

Sep 12 1999 TN, Nashville. Hot Wheels Diecast Collectors Show. Super 8 Motel, I-24 & Harding Place. SH: 9am-4pm, T: 40. Bruce Amato, PO Box 2821, Hendersonville, TN 37077. PH: 615-824-1752.

Nov 7 1999 TN, Nashville. Hot Wheels Diecast Collectors Show. Super 8 Motel, I-24 & Harding Place. SH: 9am-4pm, T: 40. Bruce Amato, PO Box 2821, Hendersonville, TN 37077. PH: 615-824-1752.

TEXAS

Jan 9 1999 TX, Waco. Wintertime Toy & Doll Show. General Exhibit Bldg., H.O.T. Fairgrounds, 4601 Bosque Blvd. SH: 10am-4pm. Productions Unlimited, 7334 N. May Ave., Oklahoma City, OK 73116. PH: 405-810-1010 or FAX: 405-787-6873.

Jan 9 1999 TX, Arlington. Texas Toyfest. Community Ctr., 3000 Matlock off I-20. SH: 10am-4pm, T: 130-8', A: $2. Robert Maston, PO Box 152, Red Oak, TX 75154. PH: 972-617-5044.

Jan 23 1999 TX, Mesquite. Toy & Doll Show. Convention Ctr., 1700 Rodeo Dr. SH: 10am-4pm. Productions Unlimited, 7334 N. May Ave., Oklahoma City, OK 73116. PH: 405-810-1010 or FAX: 405-340-3770.

Jan 30-31 1999 TX, Houston. Collectors Show. Holiday Inn, 7787 Katy Fwy., I-10 W. at Antoine Exit. SH: 10am-4pm, T: 70, A: $3., 12 & under free. Nostalgia Promos., Andy Mingle, PH: 281-748-5154.

Feb 7 1999 TX, Houston. Barbie Goes To Houston Show. Sheraton Crown Hotel Conference Ctr., 15700 JFK Blvd. SH: 10am-4pm, A: $5., $2. under 12. Marl, PH: 941-751-6275 or Joe, PH: 213-953-6490.

Feb 13 1999 TX, New Braunfels. Hill Country Doll Show. Civic Ctr., 380 S. Seguin St. SH: 9am-4pm, A: $3., $1. children. Attic Antiquity Dolls, 192 Comal Ave., New Braunfels, TX 78130. PH: 830-608-0308.

Feb 27-28 1999 TX, Austin. Collectors Exposition. Palmer Auditorium, lower level, South First & Riverside. SH: Sat. 9am-6pm, Sun. 10am-4pm. Sally Wallace, 6702 Lexington Rd., Austin, TX 78757. PH: 512-454-9882.

Feb 27 1999 TX, Gainesville. 13th Annual North TX Farm Toy Show. Civic Ctr., 311 S. Weaver. SH: 9am-4pm. Ed Pick, Rt. 2, Box 266, Muenster, TX 76252. PH: 940-759-2876.

Feb 27-28 1999 TX, Houston. Collectors Show. Holiday Inn, 7787 Katy Fwy., I-10 W. at Antoine Exit. SH: 10am-4pm, T: 70, A: $3., 12 & under free. Nostalgia Promos., Andy Mingle, PH: 281-748-5154.

Mar 14 1999 TX, Houston. 22nd Annual Doll Show. Radisson Hotel-Hobby Airport, 9100 Gulf Fwy. at the Airport Exit I-45 S. SH: 10am-5pm, A: $3., $1.50 children. Pat Black, PH: 713-864-5229 or Linda Sieck, PH: 713-667-8700 or FAX: 713-667-3606.

Mar 27-28 1999 TX, Houston. Collectors Show. Holiday Inn, 7787 Katy Fwy., I-10 W. at Antoine Exit. SH: 10am-4pm, T: 70, A: $3., 12 & under free. Nostalgia Promos., Andy Mingle, PH: 281-748-5154.

Mar 27-28 1999 TX, Plano. Sci-Fi Expo & Toy Show. Convention Ctr., 2000 E. Spring Creek Pky. SH: Sat. 10am-6pm, Sun. 10am-4pm, T: 250. Ben Stevens, PO Box 941111, Plano, TX 75094. PH: 972-578-0213.

Apr 24-25 1999 TX, Houston. Collectors Show. Holiday Inn, 7787 Katy Fwy., I-10 W. at Antoine Exit. SH: 10am-4pm, T: 70, A: $3., 12 & under free. Nostalgia Promos., Andy Mingle, PH: 281-748-5154.

May 16 1999 TX, San Antonio. The Joe & Marl Show. Doubletree Hotel, 37 NE Loop 410. SH: 10am-3pm, A: $5., $2. under 12. Marl, PH: 941-751-6275 or Joe, PH: 213-953-6490.

May 21-23 1999 TX, Plano. Sci-Fi Expo & Toy Show. Convention Ctr., 2000 E. Spring Creek Pky. SH: Sat. 10am-6pm, Sun. 10am-4pm, T: 250. Ben Stevens, PO Box 941111, Plano, TX 75094. PH: 972-578-0213.

May 22-23 1999 TX, Houston. Collectors Show. Holiday Inn, 7787 Katy Fwy., I-10 W. at Antoine Exit. SH: 10am-4pm, T: 70, A: $3., 12 & under free. Nostalgia Promos., Andy Mingle, PH: 281-748-5154.

Jun 26-27 1999 TX, Houston. Collectors Show. Holiday Inn, 7787 Katy Fwy., I-10 W. at Antoine Exit. SH: 10am-4pm, T: 70, A: $3., 12 & under free. Nostalgia

Promos., Andy Mingle, PH: 281-748-5154.

Jul 24-25 1999 TX, Houston. Collectors Show. Holiday Inn, 7787 Katy Fwy., I-10 W. at Antoine Exit. SH: 10am-4pm, T: 70, A: $3., 12 & under free. Nostalgia Promos., Andy Mingle, PH: 281-748-5154.

Aug 28-29 1999 TX, Houston. Collectors Show. Holiday Inn, 7787 Katy Fwy., I-10 W. at Antoine Exit. SH: 10am-4pm, T: 70, A: $3., 12 & under free. Nostalgia Promos., Andy Mingle, PH: 281-748-5154.

VERMONT

Feb 21 1999 VT, Bennington. 3rd Annual Model Car, Truck & Toy Show. St. Francis Parish Ctr., corner Benmont Ave. & W. Main. SH: 11am-5pm, T: 75. Kevin, PH: 802-447-7957.

VIRGINIA

Jan 2-3 1999 VA, Chantilly. Greenberg's Great Train & Collectible Toy Show. Capital Expo Ctr., Rt. 28 & Willard Rd. SH: Sat. 11am-5pm, Sun. 11am-4pm, A: $5., $2. ages 6-12, under 6 free. Greenberg Shows, Nan Turfle, 7566 Main St., Sykesville, MD 21784. PH: 410-795-7447.

Jan 9-10 1999 VA, Virginia Beach. Greenberg's Great Train & Collectible Toy Show. Pavilion, 1000 19th St. SH: Sat. 11am-5pm, Sun. 11am-4pm, A: $5., $2. ages 6-12, under 6 free. Greenberg Shows, Nan Turfle, 7566 Main St., Sykesville, MD 21784. PH: 410-795-7447.

Mar 13-14 1999 VA, Salem-Roanoke. Toy, Train & Doll Show. Civic Ctr., I-81, Exit 141. SH: Sat. 9am-5pm, Sun. 10am-4pm, T: 400, A: $5., under 12 free. Tri-City Shows, PO Box 825, Johnson City, TN 37605. PH/FAX: 888-955-TOYS.

Oct 30 1999 VA, Winchester. Backyard Buckaroo & Western Show. Nat'l. Guard Armory, Rt. 50 & I-81. SH: 9am-4pm, T: 100-125, A: $4. John Bracken, 827 Armistrad St., Winchester, VA 22601. PH: 540-678-1177.

WASHINGTON

Jan 23 1999 WA, Fife. Antique & Collectible Toys Show. Executive Inn. SH: 10am-4pm. Western Washington Harvest of Toys, Dan Marek, PH: 253-537-3172 or Charlie Ostlund, PH: 253-863-6211.

Mar 20 1999 WA, Kent. Greater Seattle Toy Show. Commons Community Ctr., 525 4th Ave. N. SH: 10am-3pm, T: 250. Todd Aicher, PO Box 1980, Snohomish, WA 98291. PH: 425-338-5958.

Apr 3 1999 WA, Spokane. Collectors Toy Show. Interstate Fairgrounds, Broadway & Havana. SH: 10am-4pm, T: 100. Creative Productions, 22221 E. Rowan, Otis Orchards, WA 99027. PH: 509-922-2773.

Jun 19 1999 WA, Spokane. Collectors Toy Show. Interstate Fairgrounds, Broadway & Havana. SH: 10am-4pm, T: 100. Creative Productions, 22221 E. Rowan, Otis Orchards, WA 99027. PH: 509-922-2773.

Dec 4 1999 WA, Spokane. Collectors Toy Show. Interstate Fairgrounds, Broadway & Havana. SH: 10am-4pm, T: 100. Creative Productions, 22221 E. Rowan, Otis Orchards, WA 99027. PH: 509-922-2773.

WEST VIRGINIA

Feb 28 1999 WV, Glen Dale. Ohio Valley Scale Modelers Contest. John Marshall High School, 1300 Wheeling Ave. SH: 10am-4pm, T: 50. William Smith, 173 Court Ave., Moundsville, WV 26041. PH: 304-845-7829 6pm-10pm.

WISCONSIN

Jan 31 1999 WI, Milwaukee. Orphans In The Attic Doll, Toy & Bear Show. Serb Hall, 5101 W. Oklahoma Ave. SH: 10am-4pm, T: 140, A: $3.50, $1.50 ages 6-12. Marge Hansen, N96W20235 County Line Rd., Meno. Falls, WI 53051. PH: 414-255-4465.

Apr 11 1999 WI, Milwaukee. Orphans In The Attic Doll, Toy & Bear Show. Serb Hall, 5101 W. Oklahoma Ave. SH: 10am-4pm, T: 140, A: $3.50. Marge Hansen, N96W20235 County Line Rd., Meno. Falls, WI 53051. PH: 414-255-4465.

Jul 9-11 1999 WI, Iola. Toy Barn Show. 700 E. State St. Bob Zurko, 211 W. Green Bay St., Shawano, WI 54166. PH: 715-526-9769.

CANADA

Jan 31 1999 ON, Mississauga. 4th Annual Toy Show. Canadian-German Club Hansa, 6650 Hurontario St. SH: 9:30am-3pm, T: 50, A: $5. Ron, PH: 416-743-2147, 219-TOYS, Dave, PH: 905-845-1824 or Claus, PH: 519-570-3120.

FOREIGN

Jan 14 1999 ENGLAND, Orpington South London. Toy & Model Collectors Fair.

Crofton Halls, Crofton Rd. SH: 7pm-10pm. Toyman Fairs, PO Box 66, Waltham Cross, Herts EN7 6NA ENGLAND. PH: 01 992 620376 or FAX: 01 992 309045.

Feb 7 1999 ITALY, Modena. 11a Mostra Mercato Del Giocattolo. Polispostiva San Faustino, Via Wiligelmo N. 72. SH: 10am-6pm, T: 150. Arci Idee, PH: 059-35.95.13 or FAX: 059-34.50.93.

Feb 11 1999 ENGLAND, Orpington South London. Toy & Model Collectors Fair. Crofton Halls, Crofton Rd. SH: 7pm-10pm. Toyman Fairs, PO Box 66, Waltham Cross, Herts EN7 6NA ENGLAND. PH: 01 992 620376 or FAX: 01 992 309045.

Mar 7 1999 ENGLAND, Rickmansworth Herts. Great Train Collectors Fairs & Shows. Watersmeet, High St. (NW London). SH: 11am-4pm. Toyman Fairs, PO Box 66, Waltham Cross, Herts EN7 6NA ENGLAND. PH: 01 992 620376 or FAX: 01 992 309045.

Apr 8 1999 ENGLAND, Orpington South London. Toy & Model Collectors Fair. Crofton Halls, Crofton Rd. SH: 7pm-10pm. Toyman Fairs, PO Box 66, Waltham Cross, Herts EN7 6NA ENGLAND. PH: 01 992 620376 or FAX: 01 992 309045.

Jul 15 1999 ENGLAND, Orpington South London. Toy & Model Collectors Fair. Crofton Halls, Crofton Rd. SH: 7pm-10pm. Toyman Fairs, PO Box 66, Waltham Cross, Herts EN7 6NA ENGLAND. PH: 01 992 620376 or FAX: 01 992 309045.

Sep 9 1999 ENGLAND, Orpington South London. Toy & Model Collectors Fair. Crofton Halls, Crofton Rd. SH: 7pm-10pm. Toyman Fairs, PO Box 66, Waltham Cross, Herts EN7 6NA ENGLAND. PH: 01 992 620376 or FAX: 01 992 309045.

Sep 12 1999 ENGLAND, Rickmansworth Herts. Great Train Collectors Fairs & Shows. Watersmeet, High St. (NW London). SH: 11am-4pm. Toyman Fairs, PO Box 66, Waltham Cross, Herts EN7 6NA ENGLAND. PH: 01 992 620376 or FAX: 01 992 309045.

Nov 18 1999 ENGLAND, Orpington South London. Toy & Model Collectors Fair. Crofton Halls, Crofton Rd. SH: 7pm-10pm. Toyman Fairs, PO Box 66, Waltham Cross, Herts EN7 6NA ENGLAND. PH: 01 992 620376 or FAX: 01 992 309045.

Nov 28 1999 ENGLAND, Rickmansworth Herts. Classic Toy Fair. Watersmeet, High St. (NW London). Toyman Fairs, PO Box 66, Waltham Cross, Herts EN7 6NA ENGLAND. PH: 01 992 620376 or FAX: 01 992 309045. PH: 01 992 620376 or FAX: 01 992 309045.

2∞

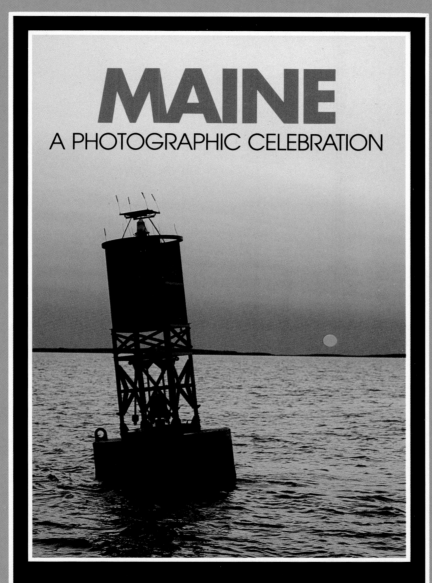

MAINE
A PHOTOGRAPHIC CELEBRATION

compiled by the staff of
American Geographic Publishing

RON SANFORD

Above: *A Maine feast near Bar Harbor.*

Title page: *On Linekin Bay.* NORMAN E. EGGERT

Front cover: *Stonington.* CHRISTIAN HEEB

Back cover, top: *A 1917 carriage bridge over the Jordan River, Acadia National Park.*
GLENN VAN NIMWEGEN
Bottom left: *Lobstering gear on Bailey Island.*
DICK DIETRICH
Right: *Atlantic puffin.* TED LEVIN

DICK DIETRICH

ISBN 0-938314-70-X

© 1989 American Geographic Publishing
P.O. Box 5630, Helena, MT 59604
(406) 443-2842

William A. Cordingley, Chairman
Rick Graetz, Publisher & CEO
Mark O. Thompson, Director of Publications
Barbara Fifer, Production Manager

Design by Linda Collins
Printed in Korea by Dong-A Printing through Codra
    Enterprises, Torrance, California

2

Above: Near Portland.
Right: Pulling oxen at the Union Fair.

3

Right: *In Bar Harbor.*
Above: *A litter of boats? Camden.*

Facing page: *A foggy day on Mt. Desert Island highlights the brilliance of Thuya Gardens' offerings.*

THUYA LODGE

Mt. Desert Island scenes:
Left: *The somber mood of twilight at low tide on a cool October evening.*
Above: *Painstaking handwork offers a welcome.*

Above: *The first touch of autumn.*

Facing page: *Sunrise over Bar I, Bar Harbor.*

Overleaf: *Glacier-carved granite shoreline and Otter Cliffs, Acadia National Park.*

FREDERICK D. ATWOOD

GLENN VAN NIMWEGEN

13

Left: *A refreshing pause at Long Pond along Route 4 near Rangely.*
Above: *Valued more than ever for their lower-fat milk, a herd of Holsteins at work near New Sharon.*

JEFF GREENBERG

15

Right: *October-kissed blueberry and bracken fern, Acadia National Park.*
Above: *Big "A" Falls on the West Branch of the Penobscot River.*

Facing page: *Camden Hills.*

17

Above: *In the Moose River area along Route 201.*

Facing page: *An inviting front porch and typically Maine architecture.*

Overleaf: *Head Beach at Small Point.*

18

20

Left: *Soft-shelled clams for dinner tonight!*
Above: *Brightening the Harbor Bridge in Rockport.*

Facing page: *The pipers are pipin' at Damariscotta's Scottish Festival.*

ERNEST J. LARSEN, JR.

23

Left: *Early morning at Mt. Katahdin, Maine's highest point, in Baxter State Park.*
Above: *Going once, going twice…at Damriscotta.*

TED LEVIN

Right: Red trillium.
Above: The clear colors of a
coastal farm.

Facing page: The capitol,
Augusta.

GLENN VAN NIMWEGEN

27

Left: *Sunset and low tide at Popham Beach State Park, Maine's longest undisturbed sand beach.*
Above: *Autumn textures, Acadia National Park.*

29

Right: *The black-capped chickadee, Maine's state bird.*
Above: *Can't you tell it's Maine? At Biddeford.*

Facing page: *Life is teeming below the surface of this tidepool at Cape Neddick Lighthouse.*

Overleaf: *Bustling Portland, Maine's largest city.*

32

33

Above: Bull moose.

Facing page: View of Penobscot Mountain across the lily pads of Long Pond, Acadia National Park.

Above: Lobster buoys and modern steel lobsterpots.
Right: The Bass Harbor Light.

34

GEORGE WUERTHNER

Blueberries:
Right: Harvesting in August.
Above: Barrens wearing autumn color on Caterpillar Hill near Sargentville.

Facing page:
Who can say more about Maine's delicious berries?

Overleaf: Protected salt marsh at Drake Island on Wells National Estuary.

RON SANFORD

37

Left: Periwinkles on rockweed exposed by low tide.
Above: Tragic reminders on the Sheepscot River at Wiscasset.

Facing page: Classic church architecture at York.

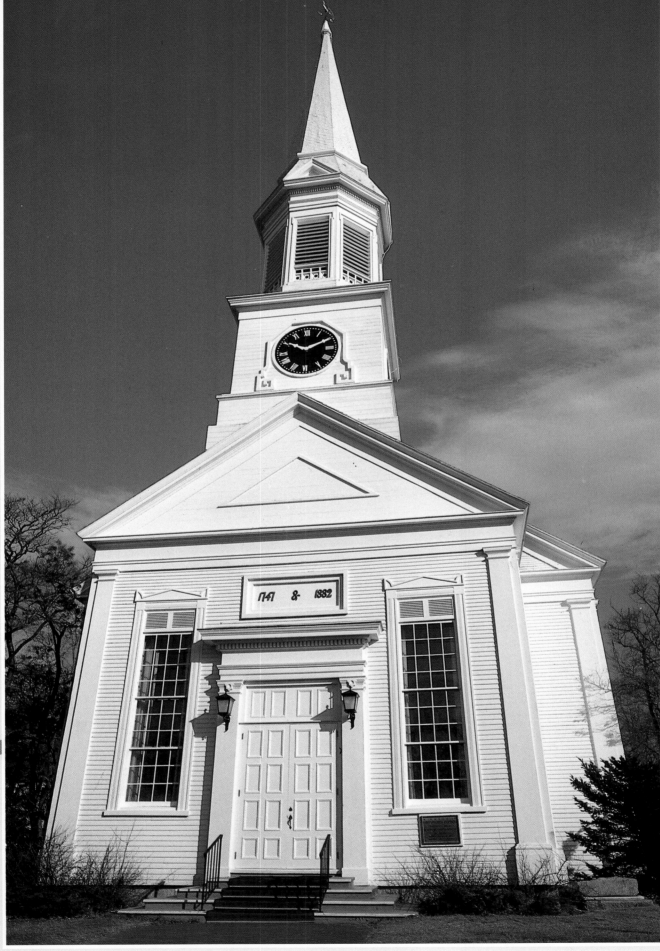

JEFF GREENBERG

43

Left: "The Beehive" at Acadia National Park.
Above: Harbor seals off Mt. Desert Island.

45

Acadia National Park scenes:
Above: Sunset barely pierces the fog to silhouette spruce trees
at Otter Point.

Facing page: The rocky terraces of an unnamed stream on
Cadillac Mountain.

46

Left: Not bad salmon fishing today.
Above: Stonington, on the Downeast coast, is truly a working fishing village.

Facing page: Fort William Henry near New Harbor.

JACK OLSON

47

JEFF GREENBERG

TED LEVIN

Right: This screech owl has had a successful hunt.
Above: Fort Popham's name honors the Englishman who sponsored one of the first attempts to establish a permanent colony nearby.
Facing page: The sun greets St. George State Park east of Augusta.

49

Above: Burchberry blossom.
Right: Rhodora and lichen-covered granite.

Overleaf: Pemaquid Point Lighthouse, one of Maine's most picturesque.

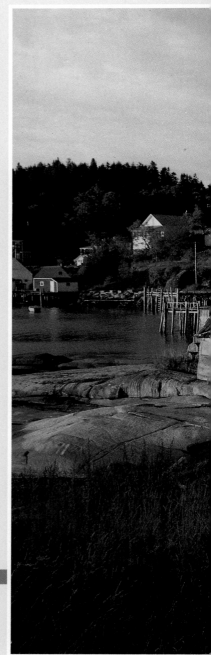

Above: Harvest time.
Right: It is easy to see why Monhegan Island is a mecca for both tourists and artists.

In Baxter State Park:
Above: *Double Top from Grassy Pond.*
Facing page: *Big Niagara Falls on Nesowadnehunk Stream.*

Right: Whitetailed buck in velvet, Baxter State Park.
Above: Auburn.

Facing page: Birch.

59

60

Above: *An antiquer's heaven.*
Right: *Basalt and granite exposed at low tide, Cobscook Bay State Park.*

Overleaf: *Camden Harbor.* DICK DIETRICH

Above: *Experiencing the colonial lifestyle at Norland's Living History Center.*

Facing page: *Quoddy Head Light.*

DIANE ENSIGN

67

Above: *Autumn splendor near New Castle.*
Left: *A quiet day at the Old Mill, Newfield.*

Right: Canada geese.
Above: Dawn mist on Whiting Narrows, upper Cobscook Bay.

Facing page: A sure sign of frosty nights.

Above: Cedar wax-wing.
Right: Morning light on Castine's harbor.

Left: Sad ending for a lovely Victorian home.
Above: At Old Orchard Beach.

Facing page: Screw Augur Falls, Grafton Notch, offers hikers cool respite on an August afternoon.

Overleaf: The Portland Head Light.

77

Above: *Along the Appalachian Trail, at the headwaters of Sandy River in the Longfellow Mountains.*

Facing page, top: *Andover's covered bridge.*
Bottom: *Summer on Mt. Desert Island.*

Above: The Coast Guard's Life Saving Station on Damariscove Island dates from 1896.

Facing page, top: Between Augusta and Wiscasset on State Road 7.
Bottom: One of these creatures is enjoying the fair at Windsor.

JEFF GREENBERG

NORMAN E. EGGERT

Above and right: *The joys of Atlantic salmon.*

Facing page: *The Dew Drop is typical of a well used Maine camp.*

Above: *Coltsfoot.*
Right: *Of course winter doesn't mean the end of fishing season.*

Overleaf: *Ogunquit.*

CHRISTIAN HEEB

Left: Autumn reflections along the Androscoggin.
Above: Acadia National Park abloom.

Above: *From the beginning, logging has been a major part of Maine's economy, as can be seen at Rumford.*

Facing page: *Fort McClary guards the New Hampshire border at the mouth of the Piscataqua River.*

Above: *Newagen in southern Maine.*

Facing page, top: *Polished by the waves on Schootic Peninsula.*
Bottom: *The venerable art of cooperage lives on.*

95

Right: A Brunswick welcome.
Above: York Beach on a July afternoon brings out the sunbathers.

Facing page: Condensing fog decorates a spider's handiwork.

Overleaf: Boothbay Harbor at day's end.